T0405577

Dragonfly-watching sites of Singapore
JOHOR BAHRU
MALAYSIA
STRAITS OF JOHOR
Sungei Buloh Wetland Reserve
Admiralty Park
Kranji Marshes
Sarimbun Reservoir
Kranji Reservoir
Murai Reservoir
Poyan Reservoir
Tengah Reservoir
Ulu Sembawang
Upper Seletar Reservoir
Lower Seletar Reservoir
SELETAR
Coney Island
Pulau Ubin
Pulau Tekong
Central Catchment Nature Reserve
Thomson Nature Park
Lower Peirce Reservoir
Chestnut Nature Park
Upper Peirce Reservoir
Dairy Farm Nature Park
BUKIT TIMAH
Bukit Timah Nature Reserve
Rifle Range Nature Park
MacRitchie Reservoir
Windsor Nature Park
Bishan-Ang Mo Kio Park
Bukit Brown
SERANGOON
Pasir Ris Park
CHANGI
CHANGI INTERNATIONAL AIRPORT
Jurong Lake
SINGAPORE
Singapore Botanic Gardens
QUEENSTOWN
CITY
East Coast Park
West Coast Park
Kent Ridge Park
Labrador Nature Reserve
Jurong Island
Sentosa Island
Pulau Bukom
Kusu Island
Sisters Islands
St. John's Island
STRAITS OF SINGAPORE
Pulau Sudong
Pulau Semakau
Pulau Pawai
Pulau Senang
INDONESIA
N
0
5 km
0
5 miles

A PHOTOGRAPHIC FIELD GUIDE TO THE

DRAGONFLIES & DAMSELFLIES OF SINGAPORE

A PHOTOGRAPHIC FIELD GUIDE TO THE DRAGONFLIES & DAMSELFLIES OF SINGAPORE

Robin W. J. Ngiam
& Marcus F. C. Ng

JOHN BEAUFOY PUBLISHING

Robin Ngiam dedicates this book to twin brothers Jian Yuan and Jian Hao: may the dragonflies always bring you joy and good health.

First published in the United Kingdom in 2022 by John Beaufoy Publishing,
11 Blenheim Court, 316 Woodstock Road, Oxford OX2 7NS, England
www.johnbeaufoy.com

10 9 8 7 6 5 4 3 2 1

Photo credits
Front cover: Male Scarlet Pygmy in obelisk © Marcus Ng.
Spine: Male Malayan Spineleg © Tang Hung Bun.
Back cover, top to bottom: Male Forktail © Marcus Ng, Male Fiery Coraltail © Tang Hung Bun, Emperor larva © Robin Ngiam.
Title page: Blue Midgets in wheel © Marcus Ng.

ISBN 978-1-912081-40-0

Edited and indexed by Krystyna Mayer
Designed by Gulmohur
Cartography by William Smuts
Project management by Rosemary Wilkinson

Printed and bound in Malaysia by Times Offset (M) Sdn Bhd

Contents

Marcus Ng

Male Crimson Dropwing at Windsor Nature Park.

Foreword

As a nature-lover, I am very happy that Singapore's new ambition is to be a City in Nature.

In spite of our small size and urbanization, nature flourishes in Singapore. We should treasure our natural heritage as much as we treasure our historical, cultural, artistic and architectural heritages.

I have been fascinated by dragonflies since I was a child. They are colourful and beautiful. They can be found everywhere in Singapore, especially near bodies of water.

I was very inspired by the book on our dragonflies, published in 2010, by our dragonfly guru, Tang Hung Bun. I have also seen, with admiration, the magnificent collection of dragonflies at our Lee Kong Chian Natural History Museum.

I warmly congratulate Marcus Ng and Robin Ngiam for their comprehensive and authoritative book on the dragonflies and damselflies of Singapore. They have discovered that we have 136, not 124, species which are native to Singapore. They have also written about the 10 species which have become extinct in Singapore.

Professor Tommy Koh

Chairman, Lee Kong Chian Natural History Museum, National University of Singapore

Patron, Nature Society of Singapore

Marcus Ng

Yellow-barred Flutterer at Lower Seletar.

Introduction

Visit nearly any well-vegetated pond, stream or reservoir in Singapore, especially around midday, and you will almost certainly see a blaze of dragonflies dashing about in hot pursuit of rivals or mates, or commanding an array of perches by the water. These insects are hard to miss, given their prominent size, active behaviour and often brilliant colours. They are also welcome sights for many, as dragonflies are well known as voracious predators of mosquitoes and other small insects that most people regard as 'pests'.

Marcus Ng

A Common Flangetail at Windsor Nature Park.

Dragonflies abound in Singapore, especially in the island's nature reserves and nature parks, which harbour ideal habitats for adults as well as larvae in the form of freshwater ponds, streams, marshes and swamps. Such wetlands, especially those located within or next to forests with dense canopies, are home to many dragonfly species that can be found nowhere else on the island. Mangroves – tidal forests that line parts of Singapore's coastline and offshore islands such as Pulau Ubin – also offer refuge to a number of dragonflies adapted to estuarine or brackish water habitats.

Beyond the forests, many of Singapore's dragonfly species can also be found in the built-up portions of the island. Artificial ponds and other man-made waterbodies in urban parks, even in the heart of the city, such as Fort Canning Park and Gardens by the Bay, support Common Parasols, Common Scarlets, Blue Dashers, Blue Sprites, Common Bluetails and other species that have adapted to human-modified environments. Wandering Gliders and Yellow-barred Flutterers often gather in feeding swarms over fields, parks and other open green spaces within the island's many housing estates. Even concrete drains and canals, despite their somewhat polluted water and lack of fringing vegetation, provide hunting and breeding grounds for tolerant species such as the Scarlet Skimmer, Blue Percher, Common Amberwing and White-barred Duskhawk.

All in all, Singapore is home to a remarkable 136 species of dragonfly, despite being a tiny island nation covering just 728.6km^2, dwarfed by New York City, USA, and less than half the area of Greater London, UK. By comparison, there are 56 or so dragonfly species in the British Isles, and 143 species in the whole of Europe. Hong Kong, to which Singapore is often compared, has 131 species.

Singapore's line-up of dragonflies includes the heaviest living dragonfly, and also one of the world's largest, the forest-dwelling Giant Hawker, which has a wingspan in excess of 16cm. The

Marcus Ng

The Spear-tailed Duskhawker is a large green dragonfly that breeds in swampy forest pools.

Marcus Ng

A Bronze Flutterer (left) and Crimson Dropwing bask at the Singapore Botanic Gardens' Symphony Lake.

island is also home to one of the world's smallest true dragonflies, the Scarlet Pygmy, a *chilli padi* (a tiny but very piquant red chilli in local Malay parlance) of an insect with a wingspan not exceeding 30mm, as well as the even tinier Dwarf Wisp, a damselfly that must rank among the world's smallest odonates.

There are no dragonflies endemic to Singapore, though. One species, the Singapore Shadowdamsel, was once thought to be restricted to the island, but has since been recorded in various locations in neighbouring Malaysia. It is no surprise that all of Singapore's dragonfly species also occur in the Malay Peninsula, which has more than 250 species (Malaysia, including the Bornean states of Sabah and Sarawak, has more than 400 dragonfly species in total). Singapore also shares at least 93 dragonfly species with Borneo, which is home to more than 300 species, many endemic to that island.

Marcus Ng

Common Parasols at a pond in the Singapore Botanic Gardens.

Singapore's rich native biodiversity, including its dragonflies, owes much to its stable equatorial climate, which is warm and humid throughout the year, as well as its location within Sundaland, a biogeographical region that includes the Malay Peninsula and parts of island Southeast Asia, notably Sumatra, Java, Borneo and Palawan. Much of Sundaland was a single land mass covered by vast rainforests and river floodplains until recently, geologically speaking. This rich and multi-faceted physical environment, coupled with periodic sea-level changes that fragmented Sundaland and isolated its animal and plant populations, fostered high rates of speciation in many groups of plants and animals, including dragonflies.

Sadly, 10 dragonfly species have become locally extinct in Singapore, although they still occur elsewhere in the region. These extirpations came in the wake of extensive deforestation and habitat loss over the past two centuries. Much of this destruction began in the early nineteenth

Chris Ang

Female Giant Hawker, one of the world's largest dragonflies, in Singapore's central forest reserve.

century, after the island was colonized by the British in 1819. Under colonial rule, vast tracts of Singapore's interior were occupied by itinerant planters who converted much of the island's original forest cover into gambier and pepper estates.

Later, in the late nineteenth and early twentieth centuries, these pioneering estates were replaced by pineapple and rubber plantations, which covered much of the island beyond its urban centres. Only steep, inaccessible areas such as Bukit Timah Hill (a pleonasm, as *bukit* means 'hill' in Malay, Singapore's national language) and uncultivable wetlands such as Nee Soon Swamp Forest were left relatively unscathed, surviving to the present as rare tracts of primary vegetation in a sea of secondary and urban landscapes. That said, the habitats that remain, both natural and modified, still harbour a diversity of dragonflies that will delight naturalists.

Why do people like us find dragonflies interesting? Along with butterflies, dragonflies are probably the most conspicuous insects that people regularly encounter, and even welcome. Dragonflies are harmless to humans, often brightly coloured and have fascinating behaviour, especially when they are hunting, patrolling and defending their territories, or engaged in courtship and mating. They are also important components of natural ecosystems and food chains. All adult dragonflies feed on smaller insects such as mosquitoes, midges and other flies; in turn, they serve as prey for birds, frogs, lizards, spiders and other creatures that share their habitats. Their larvae are also voracious predators that ambush and capture small fish, tadpoles and aquatic insects such as mosquito larvae.

Dragonflies additionally serve as useful indicators of water and habitat quality. Many species require clear and unpolluted water in order to breed, while certain others are stenotopic, meaning that they can only survive in very specific habitats such as freshwater swamp forests, mangroves, or rainforests with closed canopies or that have suffered little human disturbance. Their presence, abundance and diversity thus help us to assess the health and status of natural areas, which also support myriad other wildlife, as well as providing ecosystem services essential to our physical and mental well-being.

This book is a photographic guide to all 136 dragonfly species of Singapore, including the 10 species that have become locally extinct. It also provides basic information on the biology of dragonflies, including their anatomy, behaviour and life cycle. Sites of interest for dragonfly

Marcus Ng

The Blue-spotted Flatwing is a forest-dependent species of damselfly that perches with its wings held open.

watching in Singapore, with some tips on observing and photographing these insects, are also covered, along with an overview of dragonfly research and conservation in Singapore.

We hope that the information that follows will help you to identify the dragonflies you encounter across the island, better understand their behaviour and habitat requirements, and above all, treasure the beauty of these flying gems and support their conservation.

What is a Dragonfly?

Dragonflies are insects, like beetles, butterflies and other familiar 'bugs' – the 'true' bugs, which include the cicadas, planthoppers and stink bugs, belong to an order of their own called Hemiptera. Dragonflies belong to a taxonomic order called Odonata, a name coined in 1793 by Johan Christian Fabricius, a Danish entomologist. The name of the order is derived from *odont*, a Greek term meaning 'toothed', and refers to a pair of strong mandibles, or 'jaws', that dragonflies use to reduce prey into an edible pulp. The collective term 'odonates' (never capitalized) is often used when referring to insects in the order Odonata.

The order Odonata consists of three suborders: the damselflies or Zygoptera, the so-called true dragonflies or Anisoptera, and a very small third suborder, Anisozygoptera, which some entomologists consider to belong within Anisoptera. Anisozygoptera is regarded as a relict or 'primitive' (little-changed) group of insects that combine the traits of true dragonflies and damselflies, and contains just three or four species in a single genus, *Epiophlebia*, restricted to montane forest streams in Japan, China and the Himalayas.

Marcus Ng

The Scarlet Pygmy, a minuscule true dragonfly that is smaller than many damselflies.

Marcus Ng

The Common Scarlet, a fairly typical true dragonfly. This species is common in parks and ponds all over Singapore.

True Dragonflies (Anisoptera) & Damselflies (Zygoptera)

In this book, the collective term 'dragonfly' is used to refer to all odonates in general. Where there is a need to distinguish between the two main suborders, members of Anisoptera are referred to as 'true dragonflies', or anisopterans, while members of Zygoptera are termed 'damselflies', or zygopterans.

People often think that true dragonflies are always larger than damselflies, and invariably perch with their wings spread open. It is certainly true that anisopterans are generally bigger and more robustly built than zygopterans – however, exceptions exist. In Singapore, the Scarlet Pygmy and Pixie are true dragonflies that are dwarfed by damselflies such as the Common Flashwing and Dryad. Moreover, some groups of damselflies, such as the spreadwings (family Lestidae) and flatwings (family Argiolestidae), rest with their wings held open like true dragonflies, while a few species of true dragonfly, such as Australia's shutwings (family Cordulephyidae), rest with closed wings.

The easiest way to tell apart a true dragonfly from a damselfly is by the wings. In Greek, *zygos* means 'paired', or 'equal', while *aniso* means 'unequal', or 'dissimilar', and *pteron* refers to 'wing'. Hence, zygopterans are defined as odonates with forewings and hindwings that are similar to each other in both size and shape, although each pair of wings may have different colour patterns.

In contrast, anisopterans have hindwings that are usually larger and often much broader than the forewings, which allow them to glide and soar over ponds, fields and forests. Also, true dragonflies have relatively larger heads, which are dominated by exceptionally big eyes that touch at some point, or in the case of the clubtails (family Gomphidae), are separated by a gap less than the diameter of the eye. In contrast, all damselflies have small, delicately built heads, with eyes that are separated by a gap greater than the diameter of the eye.

Marcus Ng

Female Common Bluetail, a fairly typical and common pond damselfly.

Marcus Ng

Female Common Flashwing. This forest-dwelling damselfly is larger than many common true dragonflies.

Marcus Ng

The Blue-spotted Flatwing is a damselfly that holds its wings open like a true dragonfly.

Tang Hung Bun

The Fiery Gem belongs to an unusual family of damselflies with abdomens that are shorter than the wings.

Damian Pinguey

Female Epiophlebia superstes *(Selys, 1889) ovipositing in a mountain stream in Japan. This unusual insect has the body of a true dragonfly but damselfly-like wings.*

Despite the above-mentioned differences between the two main suborders, all dragonflies share a number of basic anatomical features. These include a distinct head with two large compound eyes and strongly toothed mandibles; a compact but strongly built thorax bearing six well-bristled legs and four very long wings, which may be clasped together over the abdomen or held open when the insect is perched; and a long, thin abdomen.

Head The compound eyes that envelope much of the head offer the insect near 360-degree vision, as anyone attempting to sneak up on a dragonfly from the back can attest. The eyes are also often multicoloured. This is due to the presence of various photosensitive pigments called opsins that absorb different wavelengths in the light spectrum and give dragonflies the ability to see a wide range of colours, thereby enhancing their visual acuity. Additionally, near the top of the head, or vertex, there are three simple eyes or ocelli in a triangular arrangement, which are thought to help orientate the insect during flight. The antennae, which are very short, unlike those of most other insect groups, are used for smell and also for flight orientation.

Between the eyes, the 'face' of the dragonfly consists of two sections called the frons and clypeus, which may feature distinctive colours, for instance, in the grenadiers and Bombardier. Below the clypeus lie the mouthparts, which consist of a labrum ('upper lip'), a pair of mandibles ('jaws') and a labium ('lower lip'). The mandibles have strong 'teeth' that are able to crush the exoskeleton of insects that form their prey. Dragonflies may attempt to bite people when badly handled, but except for the larger hawkers, their mouthparts are not likely to pierce human skin. In any case, we do not recommend handling live dragonflies, unless you are taking part in a scientific collection or survey.

Thorax This is divided into a small, neck-like prothorax bearing the first pair of legs, and a boxy synthorax consisting of a fused mesothorax and metathorax, which bear the second and third pairs of legs respectively, as well as the wings. Unusually among insects, the odonate synthorax is heavily 'tilted', so that the legs are all 'bunched up' together towards the front. This sharp angling of the synthorax allows a dragonfly to extend its bristly limbs to form a dense 'basket' that traps prey during aerial hunts.

Given this acute forwards angling of the legs, many dragonflies usually land on a vertical or

Head and thorax of a Grenadier, showing the compound eyes, ocelli, frons, clypeus, labium, mandible, labrum and major leg segments.

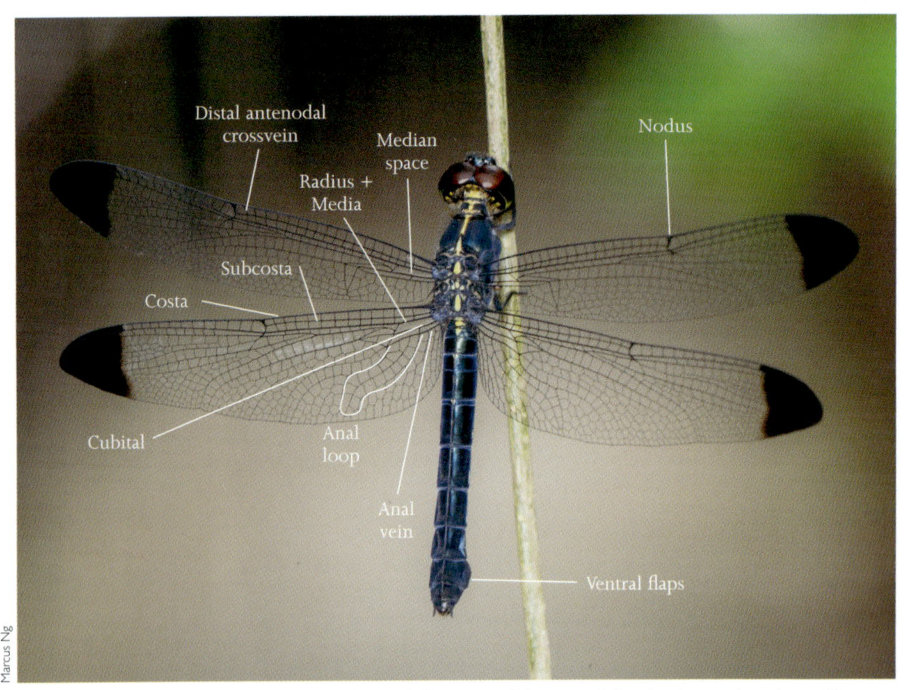

Dorsal view of a female Dark-tipped Forest Skimmer, showing the key sections of the wing, including the primary veins, and ventral flaps on abdominal segment 8.

slanted perch, which allows them to keep their bodies somewhat horizontal at rest. On a level surface, most dragonflies, except for clubtails, which have relatively short legs, have to adopt a sprawling position or settle with their abdomens pointed upwards. There are also a number of dragonflies, notably the hawkers (Aeshnidae), cruisers (Macromiidae) and certain highly aerial skimmers (Libellulidae), such as the Wandering Glider and White-barred Duskhawk, which invariably perch in a hanging position, clinging to a branch or leaf with their abdomens pointing downwards.

Wings The synthorax also bears two pairs of wings, which propel a dragonfly through the air. The next few paragraphs cover some key features of dragonfly wings and veer into technical detail, but a basic grasp of these features can be immensely helpful in identifying and distinguishing various dragonfly families and species.

The wings may be stalked, with a distinctively narrower section called the petiole at their bases, or unstalked. All true dragonflies have unstalked wings. Among damselflies, members of all locally occurring families have stalked wings, except for the satinwings (Euphaeidae) and demoiselles (Calopterygidae).

Another important wing feature is the pterostigma, a pigmented and weighted cell located along the leading edge of each wing near the tip. Present in all true dragonflies and most damselflies, the pterostigmata are thought to play a role in flight stabilization. Their shape and colours are diagnostic for certain species, such as the Malayan Grisette and Dryad.

Although they have internal differences in their venation, the wings of true dragonflies and damselflies share certain traits. They all consist of a thin but strong and flexible membrane, which is supported by six primary wing veins – the costa, subcosta, radius, media, cubitus and anal vein – some of which branch out to form distinctive wing sectors. The venation of dragonflies, as well as other insects such as butterflies and moths, is more complex, and also contested, in reality, with a number of different terminologies used. However, for the purposes of this book, this simplified explanation should suffice.

The six primary veins and their respective sectors are linked by numerous crossveins, as well as intercalated veins near the wing-tip and rear margin, which form a dense network of cells that strengthen the wing structure. Some dragonflies, such as the adjutants and baskers, have what is called an open or reduced venation, which means that their wings have relatively fewer crossveins and intercalated veins. In contrast, those with close venation, like the demoiselles, duskhawkers and flutterers, have a much denser network of crossveins.

The costa, which forms the leading or frontal edge of the wing, is divided into two sections by a 'break' near the middle called the nodus – a reinforced section of the wing that helps to absorb physical stresses while the wing twists and turns during flight. It is also a key reference point for other features of the wing, which are referred to as antenodal if they occur before the nodus, or postnodal if they are located between the nodus and wing-tip.

A case in point is the number and nature of crossveins that run between the costa and radius (the third primary vein). The number of antenodal crossveins can be used to tell apart certain similar looking species, for instance blue skimmers in the genera *Aethriamanta*, *Brachydiplax* and *Raphismia*. In addition, the distal antenodal (the crossvein closest to the nodus) may be complete, meaning that it spans the costa and radius, or incomplete, spanning only the costa and subcosta; these differences help to separate certain similar looking genera of red skimmers such as *Orthetrum* and *Crocothemis*/*Rhodothemis*.

Among true dragonflies, the base of the hindwing may have a distinctive set of cells called the anal loop, which is formed by a branch of the anal vein. The shape of the anal loop is an important diagnostic feature, as it differs between families as well as some superficially similar species, such as the grenadiers. Some clubtails lack an anal loop altogether, while skimmers often have a well-developed, sock-shaped anal loop, which may be open or closed at its (toe) end.

The shape of the wings can offer clues to a dragonfly's behaviour and habitat. Species with very long and broad wings, such as the Pond Cruiser, Saddlebag Glider and Yellow-barred Flutterer, are adapted for highly aerial lifestyles in open areas, using their wings to glide and soar with minimal effort. Damselflies, as well as true dragonflies that live in dense forests,

Lateral view of a male Dancing Dropwing, showing the main body and leg sections, abdominal segments (numbered), anal appendages and secondary genitalia.

Lateral view of a female Blue Sprite, showing the main body and wing sections, abdominal segments (numbered) and ovipositor. This individual has several mites clinging to the underside of the synthorax.

tend to have narrower wings, which help them to better negotiate enclosed spaces where manoeuvrability is more important than high lift.

Abdomen The long, thin abdomen consists of 10 segments and acts as a counterbalance in flight. It is somewhat mirrored – an evolutionary convergence – by robber flies (Diptera: Asilidae), which, like dragonflies, are highly aerial predators of small insects.

The abdomen also contains the sexual organs, with a gonopore or genital opening on the underside of segment 8 in females, and segment 9 in males. The tip of the final segment bears two appendages called cerci. These are usually short and stout in females but often highly developed in males, which use them to grasp the female during mating. Male dragonflies also possess, near the base of the abdomen, a set of secondary genitalia to which sperm is transferred prior to insemination. The abdomen of females may have, depending on the species, an ovipositor, an extended vulvar scale or pseudo-ovipositor, or a pair of ventral flaps.

Marcus Ng

An owlfly from Vietnam. The wings are superficially similar to a true dragonfly's and even bear pterostigmata. However, no dragonfly has such long antennae.

Similar Insects

In Singapore, the only other insects that somewhat resemble dragonflies are the owlflies (Neuroptera: Ascalaphidae) and antlions (Neuroptera: Myrmeleontidae). This superficial resemblance was what led early biologists, including Carl Linnaeus, the father of modern taxonomy, to classify dragonflies under Neuroptera, an insect order that currently includes the lacewings, antlions and owlflies. It was not until the early twentieth century that entomologists began to recognize Odonata as a wholly separate order of insects.

Guek Hock Ping (Kurt)

Antlions, such as this one photographed in Malaysia, have longer antennae than any damselfly.

Owlflies can be easily distinguished from dragonflies by their long, clubbed antennae (dragonflies have very short, bristle-like antennae) and are rare in Singapore, occurring mainly in areas with undisturbed tall grass. Antlions, whose larvae are fairly common in sandy areas, where they capture ants using their pit-like traps and long mandibles, may also be mistaken for damselflies. However, adult antlions have fairly long, thickened antennae and hold their wings like a tent covering their abdomen at rest, unlike most damselflies, which clasp their wings together above the abdomen. Dragonflies in the genus *Rhyothemis*, which have particularly broad and colourful wings, may also be taken for butterflies at a distance, but the differences in body shape and behaviour are apparent close up.

Marcus Ng

The Bronze Flutterer is a small dragonfly that may be mistaken for a butterfly due to its colourful wings and rather fluttery flight.

Dragonfly Diversity & Names

The total number of known dragonfly species stands at just over 6,330 in late 2021, with around 3,200 species of damselfly and more than 3,100 species of true dragonfly. However, it is thought that there are at least 7,000–7,500 living species worldwide. This estimate seems reasonable, as many new species are still being discovered and described each year, especially in highly biodiverse yet undersurveyed regions such as Borneo, Indochina, New Guinea, Central Africa and South America, where researchers are still scraping the surface as far as dragonfly diversity is concerned.

The dragonflies in this book are identified using their scientific or 'Latin' names, as well as locally accepted common names, to help readers recognize and remember them. The local common names used are based on those established in the first book on Singapore dragonflies by Tang et al. (2010).

Every described species of animal has a unique scientific name, consisting of a generic epithet that is capitalized (for example *Pantala*) and a specific epithet (like *flavescens*). This binomial or two-part name, which is always rendered in italics, is used by scientists and naturalists to refer to a particular species (for example *Pantala flavescens*), regardless of their country of origin or language. Common names, though popular and somewhat easier to recall, can be a cause of confusion, as they differ from country to country. For instance, the globally distributed *Pantala flavescens* is variously known as the Globe Skimmer, Wandering Glider or Globe Wanderer, depending on the country of the observer.

Another example is *Tholymis tillarga*, which is known as the White-barred Duskhawk in Singapore, but called the Twister, Evening Skimmer or Coral-tailed Cloudwing elsewhere. Meanwhile, *Ischnura senegalensis* is a common and very widely distributed damselfly that is known in Singapore as the Common Bluetail. However, in other parts of the world, it is referred to as the Marsh Bluetail, Ubiquitous Bluetail, African Bluetail or Senegal Golden Dartlet. Adding to the confusion, a related European species, *Ischnura elegans*, also bears the common name Common Bluetail on account of its abundance in its native range.

Sometimes, two different though similar looking species may share a common name, which further befuddles those new to dragonflies. For instance, two dragonfly species from opposite sides of the world share the common name Blue Dasher: *Brachydiplax chalybea* from Southeast Asia, and *Pachydiplax longipennis* from North America. This results in many instances of misidentification when people search the Internet for information on smallish blue dragonflies. Readers are thus encouraged to identify dragonflies by their scientific names to be precise and avoid misidentifications.

Where known, the etymology of the scientific names of dragonflies featured in this book is outlined, as an understanding of their meanings may serve as an aid to recognizing a species. Earlier scientists from the eighteenth and nineteenth centuries usually neglected to provide an explanation for the names they coined; often, the meaning is evident, for example *hyalina*

Marcus Ng

Male Wandering Glider (Pantala flavescens), *a true dragonfly with several common names due to its worldwide distribution.*

Marcus Ng

Tholymis tillarga, *a true dragonfly with several common names, including White-barred Duskhawk, Twister, Evening Skimmer and Coral-tailed Cloudwing.*

Marcus Ng

Male Brachydiplax chalybea, *which shares the common name Blue Dasher with a different species of true dragonfly found in North America.*

for a species with clear wings, but sometimes it has to be inferred, for instance *obsolescens* for a dragonfly with bronzey wings. Many dragonfly names also contain elements such as *-aeschna*, *-diplax*, *-cnemis* and *-agrion*, which arose from accidents or quirks of taxonomic history. The origins of these elements are briefly explained in the etymology sections of the species descriptions.

Flight & Feeding

Dragonflies rival or even surpass birds in their mastery of the air. In a split second, a stationary dragonfly can take off to give chase after prey or a rival before returning to its perch to feed or rest. The speed, agility and control enjoyed by dragonflies in flight stems from their ability to move each wing independently. A dragonfly can employ different wing-stroke patterns and angles of attack depending on whether it wants to patrol its territory, accelerate towards a target, bank sharply away from an approaching predator, suddenly brake in mid-air or hover to inspect an intruder into its territory.

Many true dragonflies are honed for high-speed or sustained aerial pursuit; large hawkers are said to attain speeds of more than 50km/h. Some species, such as the Emperor, Pond Cruiser and Banded Skimmer, seem tireless as they patrol the perimeters of ponds and lakes

Tang Hung Bun

Male Fiery Gem during a display flight.

Marcus Ng

Male Collared Threadtail hovering above a forest pool.

with nary a pause during the height of the day. There are also highly aerial dragonflies, such as the Wandering and Saddlebag Gliders, that seem to spend the entire day in the air – gliding, soaring and swooping after prey with minimal effort – due to their greatly expanded wings with a low wing loading (a larger wing surface area relative to body mass) that maximizes lift.

True dragonflies, with their strong powers of flight, can readily disperse far from their breeding grounds to colonize new habitats. Certain species, such as the Yellow-barred Flutterer, Wandering Glider, Coastal Glider, Blue Percher and Dancing Dropwing, often show up in urban parks, gardens and fields some distance from water, where they may spend time foraging before dispersing again to suitable breeding sites.

Taking this further, some dragonflies are long-distance travellers that vie with the famous Monarch butterfly (*Danaus plexippus*) of North America in the furthest migrations undertaken by insects. The aptly named Wandering Glider is known to regularly cross the Indian Ocean between East Africa and South Asia, covering a distance of about 6,000km. It is also the only dragonfly to have reached Easter Island, a remote isle in the Pacific Ocean more than 3,500km from continental South America.

Most damselflies, on the other hand, have traded speed for manoeuvrability. Unlike true dragonflies, typical damselflies seldom venture into the open, preferring to perch and hunt amid vegetation by ponds and streams. Damselflies have a greater range of wing-beat angles compared to true dragonflies; this allows them to weave through dense reed beds like miniature helicopters, stop in mid-air to hover or glean prey from leaves, and even fly backwards.

Compared to true dragonflies, damselflies are less able to spread to new habitats, especially those that are confined to closed forests or specific microhabitats such as seepages and swamp forests. A few common and widespread species, such as the Common Bluetail, Blue Sprite and Variable Wisp, are often encountered in urban parks and even wayside vegetation (Marcus Ng regularly sees Blue Sprites and Common Bluetails fluttering about his eleventh-storey flat at night), suggesting an ability to disperse that belies their dainty build. Damselflies that are restricted to forests, however, are not likely to cross open, exposed areas or to survive attempts to do so. Hence, their populations may suffer from the effects of isolation and limited gene flow when forests become fragmented by roads, golf courses and other urban developments that act as barriers to dispersal. For such species, which include the Common Flashwing, Telephone

Marcus Ng

A Singapore Shadowdamsel takes flight in response to the firing of a flashgun.

Sylvan, Malayan Grisette and Will-o-wisp, dense green corridors with a fairly closed canopy and interlinked waterbodies may be necessary to aid their dispersal into suitable new territories.

At times, some species of true dragonfly gather in large swarms over a pond, field, forest clearing or hilltop. Such swarms may consist of one or more species, and typically involve dragonflies that are highly aerial in habit, such as the Yellow-barred Flutterer, Wandering Glider and Banded Skimmer. This behaviour is probably a response to an abundance of prey in the area, such as a wave of freshly emerged alates (the winged reproductive forms of ants and termites), or mating aggregations of midges and other small flies. The dragonflies take full advantage of such feeding opportunities, hovering or circling about and making mid-air darts and dives to capture food. Such swarms are harmless to people and should be regarded as a vivid example of 'pest' control by dragonflies.

Erland Refling Nielsen

A Spine-tufted Skimmer hovering above a pond in Malaysia.

Erland Refling Nielsen

A Coastal Glider extending its legs as it approaches a perch. Photo taken in Malaysia.

Marcus Ng

A Spear-tailed Duskhawker, flushed from its day-time hiding spot, hovers by a trail at Thomson Nature Park.

Dragonflies as Predators & Prey

With their mastery of the air, dragonflies have their pick of prey from the array of insects that share their habitat. Generally, damselflies and smaller true dragonflies consume minute 'true' flies (order Diptera) such as midges, gnats and mosquitoes, while larger dragonflies are able to handle more robust insects such as house flies, small moths and even small bees and wasps. A

Marcus Ng

A Variegated Green Skimmer consuming a flesh fly (Sarcophagidae).

Marcus Ng

A Blue-spotted Flatwing feeding on a large crane fly (Tipulidae).

few species appear to be more specialized; the Riverhawk, for instance, is often seen preying on butterflies.

A number of dragonflies are also known to capture and feed on other odonates: these include the Common Flangetail, Common Redbolt, Trumpet Tail, Ornate Coraltail and Variegated Green Skimmer. The latter two species are notable 'cannibals' that frequently take tenerals (immature adults) and even mature adults of other dragonfly species. In North America, a few species of large dragonfly have been seen to capture and consume hummingbirds; sunbirds, the Asian analogue of hummingbirds, are probably too large for any local odonate to handle, however.

Prey is typically captured in mid-air with the aid of the long and spiny legs, which secure the victim for processing by a pair of powerful mandibles. Small prey may be simply stuffed into the mouth and consumed in flight, while larger prey is usually brought to a perch to be chewed up. Some damselflies also hunt by gleaning prey from foliage or spiders' webs, making full use of their ability to hover and manoeuvre amid dense vegetation. Blue Sprites have been seen to forage by hovering, then ramming their legs at small webs to pluck out small prey fragments or even target resident spiders.

Marcus Ng

A Variegated Green Skimmer feeding on a Common Parasol.

Marcus Ng

A Grey Sprite feeding on an Orange-striped Threadtail.

Marcus Ng

A Trumpet Tail feeding on a young Blue-tipped Percher.

That said, dragonflies also serve as prey for myriad other creatures, including, as mentioned, fellow odonates. Other insects known to prey on dragonflies include robber flies (Diptera: Asilidae), which have venom that can overpower true dragonflies larger than the asilid. Spiders, both web-spinning species and active hunters such as jumping spiders (Araneae: Salticidae), also take their toll on dragonfly populations. Damselflies, which forage low and amid dense vegetation, are particularly prone to predation by spiders, but even fairly large true dragonflies such as the Variegated Green Skimmer and Scarlet Basker can find themselves ensnared by the webs of good-sized spiders such as the common Giant Golden Orb Weaver (*Nephila pilipes*). It is not uncommon to see gleaming strands of spider silk on the wings of dragonflies that had probably blundered into a web but with sufficient velocity to escape entrapment.

Terrestrial vertebrate predators include frogs, lizards and possibly small insectivorous mammals, while in the air, dragonflies fall prey to birds, especially aerial acrobats such as swallows, flycatchers, fantails, bee-eaters and dollarbirds, but also wetland birds such as bitterns and herons. Marcus Ng has also seen a Striated Heron (*Buteroides striatus*) catching a Blue Sprite in mid-flight, not for consumption but as bait – the maimed damselfly was dropped on the

Marcus Ng

A large robber fly has captured a Common Parasol and injected its venom into the dragonfly.

Marcus Ng

*A Heavy Jumper (*Hyllus diardi*) with a female Crimson Dropwing in her jaws.*

Marcus Ng

A Common Redbolt in the web of an orb weaver spider and her smaller mate.

water's surface to draw small fish within reach of the heron's bill.

A final note on predators: humans also include dragonflies in their diet. In the Indonesian island of Bali, where dragonflies abound at rice terraces around villages, locals use poles smeared with a sticky sap called *terap*, obtained from the trunks of *Artocarpus* trees, to capture adult dragonflies, which are then roasted over a charcoal grill or cooked in coconut milk with spices. Dragonfly larvae are also collected for consumption by rural communities in Indochina and Thailand.

Parasites

Observers with keen eyes may sometimes see tiny flies attached to the body or wings of a dragonfly. These flies are usually biting midges from the genera *Forcipomyia* and *Atrichopogon*, in which females feed on haemolymph (insect 'blood') to obtain nutrients for their maturing eggs. Affected dragonflies usually have a handful of such midges on them, which remain firmly attached even when their hosts are in flight. The impact of such parasites is little known, but may be adverse when large numbers are involved – up to 171 midges have been recorded on one individual dragonfly.

Tiny larval water mites (Hydrachnidia: Arrenuridae), visible as small, often reddish spherical bodies, are sometimes seen clinging to the bodies of dragonflies. Like the biting midges, these mites feed on haemolymph and may remain on their host for up to 20 days. They then drop off into water to continue their life cycle as free-living aquatic predators. Their relationship with the host is both parasitic, as they feed from it, and phoretic, in that they use the dragonfly as a means to disperse to new habitats. Dragonflies with very high mite loads may suffer from reduced flight capabilities and reproductive success, as the mites can damage flight muscles and cause cuticle loss.

Marcus Ng

Male Common Scarlet with a rather heavy mite load.

Marcus Ng

Two biting midges feeding on a male Ornate Coraltail.

Marcus Ng

Biting midges feeding on the wing of a Treehugger.

Dragonflies by Day & Night

Most dragonflies are diurnal, meaning that they are active by day. Being ectothermic or 'cold-blooded' animals, they need a measure of heat from the sun to help power their flight muscles for strenuous activities such as pursuing prey and courtship. Also, being highly visual creatures, most dragonflies can better spot prey, rivals and potential mates in bright sunlight. In Singapore, dragonflies tend to be most active between 9 a.m. and 4 p.m., but especially so around high noon, when ponds, streams and marshes become a whirl of activity as odonates of all sizes and colours vie for optimal perches and give chase to each other without hesitation.

During the height of the day, many dragonflies can be seen perched at the tip of a twig or leaf with their abdomens raised almost vertically, or at times angled sharply downwards. This behaviour is called obelisking, after the pillar-like structure from ancient Egypt. Obelisking minimizes the amount of body surface area that is exposed to direct sunlight and prevents a dragonfly from overheating.

Like their open country cousins, dragonflies that dwell within forests also tend to be more active on sunny days, when rays of light pierce the canopy with sufficient strength to illuminate the forest floor. However, their habits differ somewhat. Forest-dependent dragonflies seldom venture into open areas, although some species, such as the grenadiers, Dark-tipped Forest Skimmer and Collared Threadtail, may bask at forest edges, though never far from the protective canopy.

Marcus Ng

Obelisking female Scarlet Basker.

Marcus Ng

A Scarlet Pygmy feeding while in obelisk.

Many forest dragonflies bask or forage in small, sunlit clearings or at partially exposed streambanks and forest pools. However, when the sky turns cloudy and the forest floor dims, these dragonflies often ascend to the canopy, where there is more light and warmth. There are also forest species, such as the Singapore Shadowdamsel, Will-o-wisp and Blue-spotted Flatwing, which seldom if ever stray from the shadowy corners of streams and swamps that form their microhabitats. Given the highly specific habitat requirements of these and many other forest dragonflies, it is crucial to preserve the heterogeneity of local forests, as well as to minimize human disturbance, to ensure that there are sufficient intact microhabitats and a broad mix of closed and semi-open spaces to support healthy populations of these dragonflies.

Towards late afternoon, many diurnal dragonflies – both those of open habitats and those of closed forests – disappear from sight, often retreating to the canopy of nearby trees to catch the last light of day. However, a few species, such as the Blue Dasher, Common Amberwing, Common Scarlet, Crimson Dropwing, Blue Sprite and Collared Threadtail, remain active by the water or trails until early evening. During such hours, these late stayers may be joined by duskhawkers (*Gynacantha* species), duskdarters (*Zyxomma* species) and the White-barred Duskhawk, which emerge from their daytime roosts as evening draws near. The latter dragonflies are crepuscular, meaning that they hunt around dusk or very early in the morning, or sometimes when the sky gets very overcast. This behaviour, with the associated adaptations for low-light vision (crepuscular species tend to have exceptionally large, light-sensitive and greenish eyes) and thermoregulation, may have evolved to help crepuscular dragonflies avoid competition with other dragonflies, evade diurnal predators such as birds, or take advantage of the vast numbers of small insects that emerge during such hours.

There are no known truly nocturnal dragonflies, although the nighthawkers (*Heliaeschna* species) are reported to be active until some time after dusk. Some crepuscular species are attracted to light, and are wont to enter lit-up buildings and compounds, where they may fall victim to frightened inhabitants or become prey for house geckos (small lizards that hunt on walls). Come evening, most dragonflies seek out roosts in reed beds, bushes or trees, which may be close to their daytime haunts but also located some distance away. Blue Sprites, for instance, are often seen fluttering up from the water's edge towards nearby vegetation, where they roost in a vertical position, clinging to a leaf or twig with the abdomen hanging downwards. True dragonflies that typically perch in a horizontal position by day, such as the Common Scarlet, Common Parasol and Yellow-barred Flutterer, also adopt a hanging position when roosting, which may allow them to quickly evade a nocturnal predator by simply letting go of their perch.

Robin Ngiam

Roosting Crimson Dropwings.

Marcus Ng

Roosting Common Scarlets at a forest edge.

Benoit Guillon

Mass roost of Wandering Gliders in Sa Pa, Vietnam.

Roosting some distance from water may help dragonflies avoid nocturnal aquatic predators such as frogs and insectivorous fish. Elsewhere in the region, species such as the Common Chaser and Wandering Glider are known to roost communally on certain trees for up to a few months, perhaps for protection in numbers. Such mass roosts have not been encountered in Singapore, but dragonflies such as the Common Scarlet, Yellow-barred Flutterer, Common Blue Skimmer, Dancing Dropwing and Wandering Glider have been seen roosting in fairly close proximity to other individuals of the same species.

The diel – daily cycle of feeding and courtship – and roosting habits of dragonflies may be partly why ponds, lakes and urban waterways with ample and varied surrounding vegetation tend to support a greater diversity and number of dragonflies compared with sparsely planted waterbodies. Richly vegetated environments, with their diverse vegetation layers and varying microhabitats, such as reed beds, shaded pools and swampy thickets, are likely to harbour a greater variety of prey as well as provide more safe roosts and spaces, especially for tenerals (immature adults) that are not yet ready to brave their more open breeding sites.

Marcus Ng

Female Dancing Dropwing. Note the absence of secondary genitalia below abdominal segment 2 (compare with image of the male on p. 16) and the short cerci at the tip of the abdomen

Marcus Ng

Male Pond Cruiser, showing the rather acute anal angle of the hindwing.

Marcus Ng

The female Ornate Coraltail is similar in colour to the male, but can be recognized by the ovipositor at the tip of the abdomen.

Reproduction

Apart from feeding and maintaining their territories, dragonflies are likely to be seen engaging in courtship and reproductive activities at their habitats. This section explains how to tell the sexes apart and also outlines the courtship behaviour, egg-laying habits and general life cycle of dragonflies.

Sexing Dragonflies

Telling apart the sexes can be a little tricky with dragonflies. A common mistake is to regard more colourful individuals as males and duller ones as females. This rule of thumb, however, is unreliable, as many dragonflies, particularly non-territorial species, are sexually monomorphic, meaning both sexes have similar colours.

Even in sexually dimorphic species such as the Common Scarlet, Common Blue Skimmer, Crimson Dropwing and Blue Sprite, young adult males often sport duller colours and patterns that closely resemble those of the females. Only when they become sexually mature do they develop the hues that give them their common names. Adding to the confusion, many common skimmers of both sexes develop a pronounced blue-grey bloom – a waxy layer known as pruinescence – on their bodies as they age, which obscures their underlying colours and patterns.

Generally, female dragonflies are slightly larger and have markedly thicker abdomens than males, which tend to be smaller and more slender in build. Additionally, male true dragonflies from certain families – notably the clubtails (Gomphidae), cruisers (Macromiidae) and hawkers (Aeshnidae) – usually have a sharp or acute anal angle at the rear corner of the hindwing-base. Females of these dragonflies can be recognized by their rounded hindwing-base.

The best way to sex an odonate, however, is to check the reproductive parts. All female damselflies and hawkers have a fairly prominent, often blade-like ovipositor below the terminal abdominal segments, which is used to insert eggs into plant matter or soil. Females of other true dragonfly families lack true ovipositors. However, many female skimmers and clubtails have a conspicuous vulvar scale, also known as a pseudo-ovipositor, which extends from abdominal segment 8 and serves as a 'chute' for their eggs. Female skimmers that lack a vulvar scale may have a pair of ventral flaps under segment 8; these help to gather their eggs in a clump before they are released into the water.

Male dragonflies, on the other hand, have pronounced anal appendages consisting of two upper or superior appendages, also known as cerci, and either two lower or inferior appendages in the case of damselflies, or a fused single lower appendage in true dragonflies. Collectively, these appendages are known as claspers and are used to grasp the rear of the female's head, or the prothorax in the case of damselflies, to secure her in a position to mate. The claspers of some true dragonflies, notably certain clubtails and cruisers, can be greatly enlarged and resemble pincers or hooks. In contrast, the final abdominal segment of all female odonates bears just two cerci, which are usually very short, except in the case of some female hawkers, like *Oligoaeschna* species, which have bizarre racquet-like cerci, and certain skimmers such as *Pantala* and *Tramea* species.

All male dragonflies have a noticeable 'bulge' on the underside of the abdomen, positioned beneath segments 2–3 in true dragonflies and segment 2 in damselflies. This ventral protuberance is part of a set of secondary genitalia, which include the penis and a visibly hook-like structure called the hamule or hamulus. Structural differences in the male secondary genitalia can be essential in distinguishing closely related dragonfly species. Females, in contrast, lack secondary genitalia, and the base of the abdomen is smooth and unadorned.

Andromorphs

Adding to possible confusion in the sexing of dragonflies is the prevalence of andromorphs in some species. Andromorphs refer to females that resemble males of their species in colouration. Locally, this phenomenon has been observed, and may be fairly common, among certain sexually dimorphic species such as the Common Bluetail, Scarlet Basker, Sultan and Common Parasol.

Copulation is a risky as well as costly act, exposing the mating pair to predators, and consuming energy and time that could otherwise be used for feeding. Thus, it is thought that andromorph females, which suffer less harassment from males and endure fewer protracted copulations, may enjoy better survival rates and thus pass on the trait to their offspring.

Marcus Ng

Andromorph Common Bluetail.

Marcus Ng

Andromorph Common Parasol.

Courtship & Copulation

In Singapore, the only dragonflies that are known to engage in prolonged courtship displays are the demoiselles (Calopterygidae) and jewels (Chlorocyphidae). Many species from these two families are sexually dimorphic, and the males may perform elaborate displays using their often colourful wings and legs to woo a mate. The female, if unwilling, may signal her rejection by opening her wings and curving her abdomen-tip downwards, or simply flying away.

With most other local dragonflies, courtship is either very brief or non-apparent. Males of many species guard small territories by a pond or stream, chasing away rivals as they await the arrival of mature females. Females of many common species, such as the Blue Dasher, Indigo Dropwing and Pond Adjutant, are seldom seen even in suitable habitats, as they probably spend most of their time foraging in the canopy or further afield, coming to the breeding sites only when they are ready to mate. Mature females that enter a male's territory are quickly pursued and grasped by the male using his claspers. In true dragonflies, this act may be so sudden and 'violent' that an observer may think the female had come under attack, until she shows up again with her head in the grip of the male. Pairs in this position, known as being in tandem, may take flight together as they prepare for copulation.

Uniquely among insects, male dragonflies must transfer sperm from their gonopore (at abdominal segment 9) to their secondary genitalia prior to mating. Male true dragonflies perform this act, known as sperm translocation, before they attempt to court or capture a female. Male damselflies, in contrast, typically do so only after they are in tandem with the female.

Marcus Ng

Pair of Crescent Threadtails flying in tandem.

Tang Hung Bun

Male Look-alike Sprite performing sperm translocation while in tandem with the female.

The clasped female initiates insemination by bending her abdomen downwards so that her genital plates attach to the male's secondary genitalia. This copulatory position is called the wheel and, especially in damselflies, it forms a vaguely heart-like silhouette that many find endearing. Some species

Marcus Ng

Blue-spotted Flatwings in wheel, with the male grasping the female's prothorax with his anal appendages.

remain in wheel, and even fly around in this position, for just a minute or less, before the female disengages and begins to lay her eggs. Others, such as the Blue Sprite and Shorttail, may remain in this position for up to an hour or more if undisturbed.

Competition between males, which is highly evident in most habitats, does not end with a successful coupling. Male dragonflies also attempt to thwart rivals using various means of sperm displacement. Depending on the species, a male may employ different means of displacing the sperm of a previous male; these include physically removing rivals' sperm using his secondary genitalia, pushing rivals' sperm to other sites in the female's genitalia, where they cannot be used, stimulating the female to eject competing sperm, and flushing out rival sperm with his own sperm. First observed in Ebony Jewelwings (*Calopteryx maculata*) in 1979, sperm displacement has since been recorded in many dragonfly species, as well as in other animals such as beetles, earwigs, crayfish and cuttlefish.

Female dragonflies are far from being passive bystanders in this arena of reproductive one-upmanship. Some females have structural adaptations in their sperm-storage organs to counter the males' tactics. These may allow a female to select sperm from a preferred mate, or to use

Marcus Ng

Violet Sprites in tandem during oviposition.

Tang Hung Bun

Female Scarlet Skimmer ovipositing while her mate hovers close by.

sperm from a variety of partners rather than just one. There is thus a coevolutionary arms race for sperm control among and between the sexes.

Whether or not they were truly successful in displacing the sperm of rivals, many male dragonflies stay close to their mates, or even remain in tandem, during oviposition. Male damselflies from the families Coenagrionidae, Lestidae and Platycnemididae typically stay attached to their mates after copulation, often adopting a vertical posture that allows the male to fly off with the female in tow when disturbed or threatened. Some male skimmers also continue to clasp the female while she releases eggs into the water, while other species practise non-contact mate guarding, with the male hovering watchfully as the female deposits her eggs in small pools.

Egg-laying Strategies

After mating and fertilization, female dragonflies typically lay their eggs in a suitable waterbody, which may be a pond, stream or even temporary pool. In urbanized areas, female dragonflies such the Wandering Glider may sometimes mistake an oil slick or the shiny chassis of a car for a suitable habitat and carry out egg-laying motions on such materials, with fatal results for their progeny.

Among odonates, there are two main methods of egg-laying: endophytic and exophytic. Endophytic species (*endo* means 'inside' while *phyto* refers to 'plant' in Greek) deposit their eggs in living or dead plant matter, or damp soil, while exophytic species (*exo* is Greek for 'outside') lay their eggs directly into water or wet banks.

Endophytic dragonflies, which include the hawkers and nearly all damselflies, possess prominent true ovipositors near the tip of the abdomen. These complex egg-laying organs may have blade-like structures that are used to make slits in the oviposition substratum before egg laying. Endophytic dragonflies usually utilize plant matter such as a piece of wood, stem or leaf, which may be above the water, or totally submerged – in the latter case the female may immerse herself for several minutes or more. Some species may utilize damp soil, old logs or plant debris near forest pools.

In contrast, exophytic species typically release their eggs directly into water in small batches. In lieu of a true ovipositor, some female skimmers and clubtails, like the Scarlet Basker and Arthur's Clubtail, have a long vulvar scale or pseudo-ovipositor that extends from abdominal segment 8; this serves as a 'chute' or 'spout' that gathers the eggs in a clump before they are

Marcus Ng

Female Blue-spotted Flatwing using her ovipositor to insert eggs into a twig above a forest pool.

Marcus Ng

Female Sultan dipping its abdomen to release a batch of eggs into an artificial lily pond at the Botanic Gardens.

released. Some skimmers, such as the Common Chaser and Scarlet Skimmer, have a pair of ventral flaps, which serve a similar function. The female then dips her abdomen into the water as she hovers or skims over the surface, releasing the egg mass, which scatters amid aquatic plants or on to the bottom substratum. Some forest-dwelling skimmers, such as the Treehugger and Dark-tipped Forest Skimmer, oviposit by repeatedly flicking their abdomens at the water, thereby flinging droplets of water along with their eggs on to a nearby bank.

A few dragonflies, such as the Dryad and Bombardier, lay their eggs only in phytotelmata or water-filled plant cavities (*phytos* is Greek for 'plant', while *telma* means 'pond'), such as bamboo stumps, large tree holes and buttress pans. Their nymphs survive by feeding on the aquatic larvae of insects such as mosquitoes (Culicidae), midges (Chironomidae) or crane flies (Tipulidae), which also use such sites for reproduction. The survival of such specialized dragonflies is dependent on mature forests with sufficient numbers of these microhabitats. They are also particularly vulnerable to the effects of climate change, as prolonged dry spells may cause phytotelmata to dry up before the larvae can reach adulthood.

Marcus Ng

Phytotelm in the form of a water-filled stump in Rifle Range forest.

Robin Ngiam

Early instar larva of a Dingy Duskhawker.

Dragonfly Larvae

When ready, the egg hatches into a prolarva, a worm-like non-feeding stage that lasts for a few hours at most. If the egg had been placed in an overhanging branch or muddy bank, the prolarva drops down or wriggles its way into the pool or stream. Either

Tang Hung Bun

Larva of a Fiery Coraltail, showing the typical form of a damselfly larva.

Robin Ngiam

Larval Emperor looking upwards, revealing the long labium with a pair of toothed labial palps at the end.

Robin Ngiam

Larva of a Chlorogomphus *species dragonfly buried in sand (except for the eyes and mouthparts), from which it ambushes prey. Photo taken in Malaysia.*

way, the prolarva quickly moults and emerges as a second instar larva with fully developed legs and mouthparts. Also known as a nymph or naiad, the larva goes through several instars (usually 8–15, depending on the species), during which it actively feeds on other aquatic creatures, including fellow odonate larvae, while avoiding predators ranging from large fish and diving beetles (Coleoptera: Dytiscidae), to birds such as kingfishers.

The larvae have little in common with their adult forms, save their predatory habits. Additionally, unlike their often colourful parents, dragonfly larvae are typically brownish or greenish, the better to camouflage themselves against aquatic plants and debris. Damselfly larvae are typically slender, with a longish abdomen ending with two or three lobe- or leaf-like structures known as caudal gills or lamellae. In contrast, the gills of true dragonfly larvae are tucked within the rectum, which also has a muscular diaphragm that can expel water with enough force to propel the larva in a jet-like manner.

Depending on the family, the larvae of true dragonflies may have highly compact bodies (Libellulidae, Corduliidae and Macromiidae), or more elongated forms (Aeshnidae and many Gomphidae), which are never as gracile as those of damselflies.

The most striking feature of dragonfly larvae and one unique to Odonata is a highly modified labium, which is long, hinged and folded under the head when not in use. The tip of the labium is equipped with a pair of expanded labial palps that form a 'mask' over the larva's face when the labium is folded. When prey is within reach, the larva rapidly extends the labium to strike at and grasp the victim with the labial palps, which bear strong, movable hooks, before withdrawing the labium to process the prey with its mandibles.

Most dragonfly larvae hunt by ambushing prey from a hidden position, amid aquatic vegetation or leaf litter, or buried under sand or silt, but some families, such as the hawkers and spreadwings, have larvae that actively chase or even swim after their prey. The larvae of Giant Hawkers are unusual in that they are semi-terrestrial: they leave the water at night and sit with their heads partially submerged to waylay shrimps or fish that approach the surface.

Eclosion & Tenerals

Dragonflies are hemimetabolous insects like mantids, grasshoppers and true bugs. This means that their life cycle goes through an incomplete metamorphosis, without the sessile pupal stage unique to holometabolous insects such as beetles, flies, butterflies and moths. Adult dragonflies develop directly from the final instar of the aquatic larva.

After a larva has reached its final instar, it eventually stops feeding, as the mouthparts within the exoskeleton develop into the adult form, rendering the larval labium non-functional. When it is ready to eclose, or emerge as an adult, the larva leaves the water. Many species use an emergent stem or other vertical perches, although clubtails prefer flat surfaces such as a stone or sandy bank. Eclosion usually takes place before sunrise, when aerial predators are less likely to be active. The imago or adult stage emerges from a fissure at the back of the larva. Once the anterior half of the body is fully extruded, the legs stretch out to grasp a stem or even its own old larval shell, allowing the imago to pull out the rest of its abdomen. It then rests, while pumping fluids into its still-compressed wings and abdomen to expand them into their adult proportions.

The newly emerged imago is a pale, fragile creature called a teneral (from *tener*, Latin for 'delicate' or 'tender'). Teneral dragonflies lack the colours and full flight prowess of the mature adult, and often seek refuge in vegetation some distance from water to avoid predators, including their own kin – tenerals make easy prey for fellow dragonflies. Only when they have built up their strength, which may take several days, do imagos venture into their breeding habitats to hunt in the open and vie for mates. In a tropical climate like Singapore's, a typical larva may require just a few months or less to grow into an adult. On the other hand, some dragonflies in temperate regions may take years to reach maturity as they undergo diapause – a period of suspended development – during cold winter months.

Tang Hung Bun

Female Variable Sprite emerging from her larval exoskeleton.

Marcus Ng

Newly emerged female Crescent Threadtail.

Marcus Ng

Teneral male Shorttail. The pale eyes and sheen on the wing are typical of teneral dragonflies.

Marcus Ng

Young adult Blue Percher, which still has the shiny wings of a teneral.

Dragonfly Watching in Singapore

With its year-round equatorial climate and ease of transport to various nature areas, Singapore offers a considerable number of sites of interest to dragonfly watchers. Many of these locations are readily accessible via public transport, including buses, the local metro or MRT, taxis, private-hire car services or ferries, and none of them charges entry fees. For international visitors, most decent hotels are located in or close to the city centre or the east-coast suburbs, near the airport, but nature areas are usually just a bus, train or taxi ride away.

Urban Parks & the Singapore Botanic Gardens

Singapore has plenty of public parks, even in the heart of the city, such as Fort Canning Park and Istana Park, which lies in the middle of the Orchard Road shopping belt. These urban parks often have small ponds or artificial waterbodies that attract a handful of common dragonfly species. The largest of these parks, the sprawling Gardens by the Bay at the southern end of the business and tourist district, is better known for its so-called (artificial) Supertrees, but this swathe of reclaimed land contains a network of well-vegetated ponds and waterways, including the well-named Dragonfly Lake, from which more than 20 dragonflies, mostly common, open-country species, have been recorded.

Further afield, parks in the island's many public housing estates – such as Toa Payoh Town Park, Yishun Park and the adjacent Yishun Pond Park, Bishan-Ang Mo Kio Park, Punggol Park, Jurong Central Park and Jurong Lake Gardens – harbour ponds, recreated wetlands or pockets of woodland that support a greater diversity of dragonflies adapted to open country or human-modified environments. Less common species may also show up at these sites: the rare Orange-faced Sprite has been recorded in Toa Payoh Town Park, while the uncommon Fiery Coraltail was formerly established at Bishan-Ang Mo Kio Park. In 2019, a rare Spoon-tailed Duskhawker was encountered at a hospital in Yishun Town Central, which is located next to Yishun Pond Park. A study published in 2011 recorded 51 dragonfly species in 19 urban parks across Singapore; this count is likely to have increased since then, given fresh dispersals, the maturing of these habitats and new sightings by avid dragonfly watchers.

There are also public parks set within more densely wooded areas. Along the southwestern

Robin Ngiam

Pond in Toa Payoh Town Park, where at least 18 species of dragonfly, including the uncommon Orange-faced Sprite, have been recorded.

Marcus Ng

Freshwater pond at Pasir Ris Park, the habitat of the Variable Wisp, Common Chaser, Common Parasol, Scarlet Basker and Scarlet Grenadier.

coast, Kent Ridge Park and the adjacent HortPark contain a few productive ponds at the lee of a forested slope, where uncommon species such as the Scarlet Adjutant and Sapphire Flutterer regularly appear. Meanwhile, Admiralty Park (named thus because it was once part of a British naval base) in the north has a boardwalk over part of Sungei Cina, a tidal creek, where mangrove-dependent species such as the Mangrove Marshal and Arthur's Midget can be easily seen in the company of marine creatures such as mudskippers and crabs. At the northeastern corner of the island, Pasir Ris Park, Coney Island Park and Lorong Halus Wetland are fairly fruitful sites for odonates associated with mangroves and semi-open coastal habitats, such as the Mangrove Dwarf and Lined Forest Skimmer.

By far the most rewarding urban park for odonates is probably the Singapore Botanic Gardens, a UNESCO-inscribed World Heritage Site located just north of the Orchard Road shopping district. At least 40 dragonfly species – including the uncommon Bronze Flutterer and rare Dwarf Wisp – have been recorded in the Gardens, which contain several well-vegetated waterbodies such as the Swan Lake, Marsh Garden, Symphony Lake, Eco-Lake and Keppel Discovery Wetlands, which recreates the region's dwindling swamp forest habitat. At the Gardens' historic Plant House, there is a small artificial lily pond that is used by the uncommon Sultan and the Shorttail as a breeding site.

Nature Parks, Nature Reserves & Pulau Ubin

Beyond the built-up and suburban areas of the island, Singapore has a number of nature parks and nature reserves, which were established to preserve Singapore's biodiversity and remaining natural habitats. These protected areas offer a refuge for many species of native wildlife, including dragonflies, which need the shelter of a dense forest canopy, natural streams and relatively undisturbed wetlands.

The nature parks consist of secondary forests of varying age, along with modified landscapes such as ponds, parkland and disused quarries that provide avenues for outdoor recreation. Some of the nature parks also serve as 'buffer parks' that protect the core forests of the Bukit Timah and Central Catchment Nature Reserves from overcrowding and development at their edges. Straddling as they do the dense central forests and somewhat more open environments at their fringes, these nature parks offer a rich assembly of odonates, both forest-dwelling species and those associated with more exposed habitats.

Marcus Ng

Stream in Windsor Nature Park, the habitat of the Violet Sprite, Orange-striped Threadtail, Indigo Dropwing, Spine-tufted Skimmer, Tiny Sheartail, Crescent Threadtail and Blue-sided Satinwing.

Marcus Ng

Swampy forest patch next to a trail at Thomson Nature Park.

Some nature parks, such as Windsor Nature Park and Thomson Nature Park, have well-shaded, boggy patches that attract swamp-loving dragonflies like the Blue-spotted Flatwing, Collared Threadtail, Red-tailed Sprite, Will-o-wisp, Spear-tailed Duskhawker and Handsome Grenadier. Windsor Nature Park also has a well-vegetated pond near its entrance that has attracted more than 30 species, including uncommon dragonflies such as the Sultan, Scarlet Adjutant, Green-eyed Percher and Bronze Flutterer.

At Dairy Farm Nature Park – which was indeed a dairy during colonial times – an old concrete cow trough now serves as a breeding site for the Ornate Coraltail, Grenadier and Dark-tipped Forest Skimmer, along with tree frogs whose tadpoles provide prey for odonate larvae. Streamside trails in the nature parks also offer opportunities to see the Common Flashwing, Crescent Threadtail, Violet Sprite, Ris' Clubtail, Variable Sentinel, Indigo Dropwing and other species that do not occur in urban parks and waterbodies.

By far the richest nature areas in terms of local biodiversity are the forests of the Bukit Timah and Central Catchment Nature Reserves, which lie in the heart of the island. Bukit Timah Nature Reserve contains the largest contiguous tract of lowland dipterocarp primary forest in Singapore and harbours at least 55 odonate species. However, the reserve itself is not the most fruitful site for watching dragonflies (the adjacent Dairy Farm Nature Park is preferable), as the streams there tend to be steep and inaccessible, and its trails are often crowded with hikers, especially at weekends. That said, Bukit Timah offers an opportunity to spot species associated with fairly pristine rainforest, such as the Malayan Grisette, Singapore Shadowdamsel, Dryad, Emerald and Bombardier.

The Central Catchment Nature Reserve, on the other hand, offers a network of rewarding trails through a mosaic of primary and mature secondary rainforests. Forest-dwelling dragonflies such as the Treehugger, Dark-tipped Forest Skimmer, Common Flashwing, Telephone Sylvan and Grenadier are often encountered along these trails, which also cross or run past streams and swampy areas where you can spot species such as the Golden Gem, Blue-sided Satinwing, Blue-spotted Flatwing, Blue-nosed Sprite and Stream Cruiser. Jelutong Tower, a seven-storey-high tower near the centre of the reserve, is a popular observation post for birdwatchers, but also a good vantage point from which to spot clubtails and cruisers hunting over the canopy, or skimmers foraging in the treetops.

The southern portion of the central reserve is dominated by MacRitchie Reservoir, which has marshy fringes on parts of its western and southern banks. Here, a boardwalk and waterside trail brings you close to wetland species such as the Crenulated Spreadwing, Look-alike Sprite, Scarlet Pygmy, Black-tipped Percher, Green-eyed Percher and Sapphire Flutterer.

At the northeastern end of the reserve is Nee Soon Swamp Forest, a semi-flooded forest that has suffered limited human disturbance and thus shelters many species (not only odonates) that are highly demanding in terms of water quality and the surrounding forest conditions. Criss-crossed by shaded streams and swampy pools, Nee Soon Swamp Forest is home to 68 dragonfly

Robin Ngiam

Pond at Kranji Marshes, a wetland where more than 33 species of dragonfly, including the rare Plain Nighthawker, have been recorded.

species, including some that occur nowhere else on the island, such as the Malayan Spineleg, Arthur's Clubtail, Cryptic Shadesprite, Bebar Wisp, Interrupted Threadtail, Blue Sentinel and Potbellied Elf. This area is restricted to the public due to its ecological sensitivity, but some of its species may be encountered at the forest edges of Upper Seletar Reservoir Park, which abuts the northern fringes of Nee Soon Swamp Forest.

Singapore's two other nature reserves, Labrador Nature Reserve and Sungei Buloh Wetland Reserve, are also worth visiting for odonates. The former, easily reached via Labrador Park MRT station, consists of a coastal hill forest that overlooks a marshy field with a small artificial pond, at which 34 species, including the rare Wandering Wisp and uncommon Restless Demon, have been recorded.

Sungei Buloh Wetland Reserve, on the northwestern coast, consists of extensive mangroves and former prawn ponds that have become a haven for migratory birds. This reserve is better known for its avifauna and apex predators such as Estuarine Crocodiles (*Crocodylus porosus*) and Smooth Otters (*Lutrogale perspicillata*), but it is also a fairly productive site for odonates, with more than 43 species recorded to date. Near the reserve's entrances are freshwater ponds where open-country dragonflies, including the uncommon Blue Adjutant and Restless Demon, have been spotted, while the trails further in provide a chance to encounter species such as the Pond Cruiser, Mangrove Dwarf and Arrow Emperor. In the same vicinity are two nature parks that form part of Sungei Buloh's extended network – Kranji Marshes, and Mandai Mangrove and Mudflat (scheduled to open in 2022) – which harbour a good diversity of open country and mangrove-associated odonates.

The nature reserves and nature parks are bastions of Singapore's natural heritage, being 'islands' of biodiversity within a sea of urban development. To protect the fragile habitats and wildlife in these nature areas, visitors should stay on designated trails and adhere to the rules stated on public notices at these sites. Entry into the nature reserves and most nature parks at night (between 7 p.m. and 7 a.m.) is also prohibited to avoid disturbing nocturnal wildlife such as pangolins and colugos.

Although not a gazetted nature reserve, Pulau Ubin, a small, relatively undeveloped island off the northeastern coast, deserves special mention as another productive site for watching dragonflies. The island is easily reached via bumboats (small ferries) from Changi Point Ferry Terminal, which brings visitors to a coastal village with shops, eateries and temples that offer a glimpse of Singapore's rustic past.

Formerly dominated by granite quarries (*ubin* is Malay for 'granite'), rubber estates and prawn ponds, much of Pulau Ubin is now a nature area with secondary forests, mangroves and well-vegetated freshwater ponds. At least 54 species of dragonfly have been recorded in Pulau Ubin, including some that are absent or uncommon on the mainland, such as the Variable Featherlegs, Arthur's Midget, Lined Forest Skimmer, White Duskdarter and White-tipped Demon. Fruitful areas for finding dragonflies include the Sensory Trail just east of the main village, which winds through mangrove creeks and coastal woods, and Chek Jawa Wetlands at the island's eastern tip. These and other sites are easily reached by walking, renting bicycles from local shops, or hiring a van and local driver.

Dragonfly Conservation & Research

Due to their status as flagship and indicator species for freshwater and wetland habitats, dragonflies have been accorded considerable attention in a wide range of conservation projects globally. These include efforts to preserve important dragonfly habitats and hotspots. Perhaps the best example of this is Dragonfly Kingdom, a nature park near the town of Nakamura in Shikoku, Japan, which opened in 1986. Harbouring about 80 species, this was the first nature area in the world to be established explicitly for dragonflies. Since then, many other similar parks have been opened in Japan.

Apart from preserving habitats of value, dragonfly conservation efforts have involved the introduction of rare or threatened species at sites where they had been extirpated, with the hope that they will become re-established, or at suitable new locations, to expand the range of the species and mitigate the impact of possible future local extinctions. Where successful, such programmes may help enhance the resilience of rare dragonflies and avert their extinction should habitat loss or climate change result in their extirpation at certain sites. In the Czech Republic, penultimate instar larvae of the locally rare White-faced Darter (*Leucorrhinia dubia*) were introduced to artificial peaty bog pools, where they established thriving populations without substantially changing the existing dragonfly assemblage.

Marcus Ng

Mangroves at Pasir Ris Park. Such intertidal forests are the habitat of stenotopic dragonflies such as the Mangrove Dwarf, Mangrove Marshal and Arthur's Midget.

In East Asia, a number of dragonflies have become threatened with extinction. The Bekko Tombo (*Libellula angelina*), a skimmer native to China, Japan and South Korea, is classified as Critically Endangered by the International Union for Conservation of Nature (IUCN), due to the loss or degradation of its preferred habitat of old ponds with moderately rich vegetation and areas of open water. Another dragonfly found in parts of East Asia, the Mangrove Skimmer (*Orthetrum poecilops*) is regarded as Vulnerable to extinction by the IUCN, as it is dependent on mangroves and intertidal mudflats, which are rapidly being converted into coastal and aquaculture developments.

The precarious state of many mangroves in the region is also why Arthur's Midget and the Mangrove Marshal are among the local dragonflies to be evaluated by the IUCN as globally Near Threatened species, due to their highly specific habitat requirements and patchy occurrence in just 10 or so known sites across Southeast Asia. The Singapore Shadowdamsel is also considered Near Threatened by the IUCN, as this forest-dependent damselfly is known from just a handful of locations in Singapore and Johor in Malaysia. Its habitats in Singapore are fortunately located within protected areas, but in Johor it is threatened by deforestation for the development of oil palm plantations.

Other globally Near Threatened local odonates include Rebecca's Sprite, which has been recorded at less than 20 locations in the region and just one site in Singapore, and the Red-tailed Sprite, which is dependent on healthy swamp forests. The Bebar Wisp and Slender Spreadwing are designated as Data Deficient by the IUCN due to lack of knowledge about their populations, but are likely to be globally threatened as well, given their scarcity. Both these damselflies are regarded as Critically Endangered species in Singapore as they are restricted to a single location.

Among other extant dragonflies in Singapore, the clubtails (a family known for its fastidious habitat requirements) are the most threatened as a group – 10 of the 11 known native species are either locally extinct or locally threatened. The Malayan Spineleg is classified as globally Endangered by the IUCN, while the Malayan Grappletail is considered Vulnerable and the Malayan Hooktail is Near Threatened. Their known habitats in Singapore are thankfully under protection, but these dragonflies may still be susceptible to the impact of a warming climate or cascading ecological effects from other changes in their environments, such as the loss of suitable prey or the introduction of invasive species. Other anthropogenic threats include physical and chemical pollution of habitats by litter, and by commercial or industrial outflows.

In Singapore, conservation efforts that have benefited dragonflies, as well as other freshwater

Marcus Ng

Sandy streams in closed forests are vital to the survival of dragonflies such as forest-dependent clubtails.

species, include the restoration, enhancement or creation of wetland habitats, which help to enhance dragonfly abundance and diversity. The National Parks Board is the government body that manages the island's designated nature areas and biodiversity conservation programmes. Apart from taking care of Singapore's four gazetted nature reserves, the Board has converted selected nature areas into nature parks that serve as buffers to the nature reserves. These buffer parks, which help to protect the nature reserves and relieve overcrowding by providing alternative venues for outdoor recreation, also contain pockets of valuable habitats where people can see and enjoy forest-dependent dragonflies. In the near future, new nature parks that will be added to Singapore's landscape include Rifle Range Nature Park, south of Bukit Timah Nature Reserve, Mandai Mangrove and Mudflat, east of Sungei Buloh Wetland Reserve, and Lim Chu Kang Nature Park, west of Sungei Buloh.

That said, public access to nature areas is a double-edged sword. Singapore Quarry, a rehabilitated granite mine in Dairy Farm Nature Park, was formerly a stronghold of the locally uncommon Crenulated Spreadwing, but these damselflies vanished from the site after members of the public (illegally) released unwanted exotic pet fish such as Japanese koi and various cichlids into the quarry waters. The American Bullfrog (*Lithobates catesbeianus*), a voracious predator that is a common escapee from farms and restaurants or is intentionally released by 'kind-hearted' people, has been seen devouring an ovipositing Emperor at an urban pond, and represents a threat to which local dragonflies may not have adapted. The only native frog of similar size is the Malayan Giant Frog (*Limnonectes blythii*), which is nocturnal and restricted to forest streams.

The National Parks Board is also implementing species recovery plans for selected native flora and fauna. Focusing on species with small populations and very restricted distributions, these species recovery plans involve habitat enhancement and protection efforts, as well as reintroduction measures. The latter includes an ongoing effort to expand the range of the Pixie, a small dragonfly that is Endangered in Singapore, to suitable sites beyond its only known locality in the western corner of the island. The board has also undertaken comprehensive biodiversity surveys, including studies of the odonate assemblies, of key habitats such as Nee Soon Swamp Forest and Bukit Timah Nature Reserve, the results of which were published in 2018 and 2019 respectively (see Selected References, p. 337).

Research & Outreach on Dragonflies in Singapore

Dragonfly conservation efforts in Singapore, though a relatively recent development, owe much to research and documentation by various collectors and scientists over more than a century. The first known collection of local odonates was created by Alfred Russel Wallace, a British zoologist and co-discoverer of the theory of evolution by natural selection, who spent a couple of months at Bukit Timah in mid-1854. He was more enamoured by the hill's butterflies and beetles, but noted in one letter that 'of dragon-flies I have [collected] many pretty species'. Wallace is thought to have collected at least 36 species, including several that were later described as new, in Singapore. The Belgian entomologist Baron Michel Edmond de Sélys Longchamps used specimens collected by Wallace in Singapore to scientifically describe several species including the Pixie, Bombardier, Malayan Grisette, Telephone Sylvan and Fiery Gem.

From Singapore, Wallace continued his explorations elsewhere in the Malay Archipelago. At Mt Ophir (Gunung Ledang), near Melaka, he found 'hosts of new and beautiful dragon-flies' at streams in the foothills. In 1856, in Ampanam, a town in Indonesia's Lombok Island, Wallace made an odonatological observation that still rings true today:

> Every day boys were to be seen walking along the roads and by the hedges and ditches, catching dragon-flies with bird-lime. They carry a slender stick, with a few twigs at the end well anointed, so that the least touch captures the insect, whose wings are pulled off before it is consigned to a small basket. The dragon-flies are so abundant at the time of the rice flowering that thousands are soon caught in this way. The bodies are fried in oil, with onions and preserved shrimps, or sometimes alone, and are considered a great delicacy.

After Wallace, dragonfly collections from Singapore were made by other researchers. Among them was Friedrich Ris, a Swiss odonatologist after whom Ris' Clubtail and the genus *Risiophlebia* were named, who collected several species during a one-day visit on 10 April 1891. Henry Nicholas Ridley, director of the Singapore Botanic Gardens in 1888–1912, also collected dragonflies on occasion, which he sent to the British Museum (now the Natural History Museum in London).

In July 1899, British entomologist Frank Fortescue Laidlaw collected many new dragonflies in Singapore and Malaya as part of the Skeat Expedition funded by Cambridge University. Laidlaw then produced in 1902 the first list of the area's dragonflies, *On a Collection of Dragonflies made by Members of the Skeat Expedition in the Malay Peninsula in 1899–1900*.

Laidlaw followed up on this in 1931 with *A List of the Dragonflies (Odonata) of the Malay Peninsula with Descriptions of New Species*, which included the localities, where available, of specimens from Singapore. He wrote therein: 'I have attempted to note all records for Singapore as it occurs to me that such records may be of particular interest in view of the many changes which have been in progress in the Island in the last hundred years or so.' The next important overview of the region's dragonflies was Dutch entomologist Maurits Anne Lieftinck's 1954 *Handlist of Malaysian Odonata*, which also cites specimens originating in Singapore.

Many of the specimens referenced by Laidlaw were obtained from the directors of the Raffles Library and Museum, which was founded as a public institution in 1878. During the pre-war era, directors such as Karl Richard Hanitsch, John Coney Moulton, Cecil Boden Kloss and Frederick Nutter Chasen helped to enrich the museum's entomological collection, although odonates were not their main taxa of interest (Hanitsch, for instance, specialized in cockroaches). The museum's oldest dragonfly specimens are three skimmers collected in May and September 1921 – two of these by Chasen, then the museum's taxidermist. Chasen is also honoured in odonatology with *Calicnemia chaseni*, a bright red damselfly endemic to the Malay Peninsula, named after him by Laidlaw in 1928.

Raffles Library and Museum became separate institutions in 1957. In 1960, as Singapore inched towards nationhood, Raffles Museum was renamed the National Museum (now the National Museum of Singapore). Its focus then shifted from natural history in favour of national history, art and culture; hence, in 1970, the museum's entire collection of zoological specimens was transferred to the Singapore Science Centre, a newly established science education facility, which then donated the collection to the University of Singapore's Department of Zoology in 1972.

This priceless reference collection subsequently endured numerous shifts and precarious storage conditions in various facilities, before it found a permanent home in the National University of Singapore's Department of Biological Sciences in 1987. The university established the Raffles Museum of Biodiversity Research to manage the collection in 1998. With growing recognition of the importance of its collection and biodiversity research in Singapore and Southeast Asia, the university then built a dedicated facility to house the collection and a public gallery, which opened in 2015 as the Lee Kong Chian Natural History Museum.

To date, the museum contains more than a million zoological specimens belonging to at least 10,000 species. These include at least 5,000 odonate specimens (adults, larvae and exuviae), of which about 3,000 were deposited by José I. dos R. Furtado, a former professor of zoology at the University of Malaya, who studied the ecology and behaviour of Malaysian dragonflies in the 1960s and '70s. Known as the Furtado Collection, this collection

Marcus Ng

Damselfly specimens in the collection of the Lee Kong Chian Natural History Museum.

includes dragonflies from Peninsular Malaysia, Borneo, southern Thailand, the Netherlands, Germany and Canada.

Another important contributor to the museum's collection was the late Dennis Hugh Murphy, a prolific entomologist who collected some 20,000 insect specimens, including many odonates. In 1997, Murphy produced a landmark survey of the dragonflies of the Bukit Timah and Central Catchment Nature Reserves, covering 79 species, including new country records such as the Crenulated Spreadwing, White-tailed Sylvan and Blue-nosed Sprite. In 2008, Murphy, together with Y. Norma-Rashid, Cheong Loong Fah and Lua Hui Kheng, published *The Dragonflies (Odonata) of Singapore: Current Status Records and Collections of the Raffles Museum of Biodiversity Research*, which provided an annotated checklist listing 117 species for Singapore.

The national list had risen to 124 species by 2010, when Tang Hung Bun, Wang Luan Keng and Matti Hämäläinen wrote *A Photographic Guide to the Dragonflies of Singapore*. Now out of print, this was the first book to present Singapore's odonates to a general audience, and must be credited with expanding interest in the island's dragonflies among both locals and international visitors. Singapore's dragonflies were also covered in A. G. Orr's *Dragonflies of Peninsular Malaysia and Singapore* (2005), Robin W. J. Ngiam's *Dragonflies of our Parks and Gardens* (2011) and *A Field Guide to the Dragonflies of Singapore* (2012) by Lena Chow, Gan Cheong Weei and the late Tsang Kwok Choong. The former two books are also out of print, attesting to the popularity of their subject.

In the decade since Tang et al. (2010), periodic surveys, serendipitous sightings by amateur and professional naturalists, as well as delvings into the holdings of European natural history museums, have raised Singapore's dragonfly count to 136 species. Local odonatology has also been enriched by a growing pool of naturalists who have charted the distribution, life cycles and ecology of lesser known species. Many of their papers are available online as part of the Lee Kong Chian Natural History Museum's *Nature in Singapore* journal, first published in 2008, and *Singapore Biodiversity Records*, which began in 2013 and became the Biodiversity Records section of *Nature in Singapore* from 2021. The latter includes key reviews of the checklist and conservation status of Singapore's dragonflies, published in 2016 and updated in 2019, which form the basis of the present book.

The aforementioned books and other publications have helped dragonflies make inroads into the public consciousness, and there is now an active community of dragonfly watchers who share their sightings on social network platforms such as iNaturalist, a citizen science portal where wildlife observations can be uploaded as data for researchers, and the Dragonflies of Singapore Facebook group. The latter is an active online community dedicated to odonates, with regular postings and discussions on dragonflies from all over the island. Amateur naturalists have made valuable contributions to local odonatology, through sightings of species such as the White Duskdarter and White-tipped Demon that were confirmed as new country records, and by documenting the occurrence of rare species such as the Malayan Spineleg, Spoon-tailed Duskhawker, Bombardier and Wandering Wisp.

Public interest in dragonflies has also risen due to the high visibility of these insects and increased accessibility to their habitats, in the form of trails and boardwalks in nature areas, and the creation of suitable habitats in formerly barren or neglected urban areas. In the northern district of Seletar, an old pond within a former air base has been converted into a nature area called Hampstead Wetlands Park, while Rowers' Bay, formerly an undifferentiated portion of the Lower Seletar Reservoir, has been remade into a heterogenous yet accessible landscape of swales and urban wetlands. Boardwalks and pavilions installed in these areas allow visitors to go close to the water and observe the rich wildlife, including dragonflies.

A forerunner of this trend is Bishan-Ang Mo Kio Park, where a concrete storm canal, originally the upper reaches of the Kallang River, was converted into a naturalized wetland with a meandering stream and surrounding floodplain that doubles as a flood-protection measure and a habitat for aquatic creatures including birds, otters and more than 30 species of dragonfly. Bishan-Ang Mo Kio Park is also one of many sites surveyed by volunteers of Dragonfly Watch, a programme under the National Parks Board's Community in Nature initiative to encourage citizen science. Volunteers with Dragonfly Watch receive training on how to identify common dragonflies, record their observations and survey dragonflies at designated parks across Singapore. These surveys provide data on the species and abundance of dragonflies in Singapore's parks and gardens, and contribute towards improved park management and conservation measures. Further public outreach on dragonflies has been carried out by Robin Ngiam through talks and training courses aimed at generating interest and knowledge in dragonflies and their conservation.

Creating a Habitat for Dragonflies

Tang Hung Bun, lead author of the first popular guide to Singapore's odonates, remains an avid dragonfly watcher. He has also ventured into habitat restoration and permaculture, including the rehabilitation of urban gardens and waterbodies into sites that can support native biodiversity while providing food and other ecosystem services for people. Here he shares his experience of converting a formerly bare pond within a school compound into a thriving habitat for dragonflies and other wildlife.

Due to extensive urbanization, Singapore has been losing wildlife habitats over the years. Restoring wildlife habitats and ecosystems will not only safeguard our biodiversity, but also enhance the health and well-being of people living and working in urban Singapore. Some uncommon species of dragonfly do exist in the green spaces within our urban areas. The beautiful Fiery Coraltail (*Ceriagrion chaoi*) used to occur in a pond in Bishan-Ang Mo Kio Park, with a thriving population until 2012, when the pond habitat was altered.

In 2019, I had the chance of transforming a school garden into a biodiverse edible garden. The garden included a concrete koi pond, which I converted to an eco-pond with a focus on dragonflies. The concrete pond was of a decent size, about $60m^2$, but had a uniform depth of about 0.4m and a bare concrete bottom that could not support any aquatic plants. An eco-pond needs to have shallow water as well as deeper water regions to provide a diversity of microhabitats for aquatic plants and animals. While I could not increase the depth of any part of the pond, I created shallow-water regions by adding pond mud to a small corner on one side of the pond, using bricks to border up the area. I also placed large baskets containing pond mud in various parts of the pond.

Robin Ngiam

Tang Hung Bun at the dragonfly pond he created and maintains in the grounds of a school.

Floating and emergent plants were then planted in these shallow-water areas to create larval habitats for dragonflies. A variety of submerged and emergent plants was also planted all around the pond, using unglazed clay pots with a substratum of pond mud, gravel and pebbles of different sizes. The surfaces of the pebbles and the porous clay act as media for beneficial bacteria that help to recycle nutrients and maintain a balanced pond ecosystem. As the pond was already quite adequately shaded by two mature mango trees, only about a third of its surface was covered by a diversity of aquatic plants.

Submerged plants provide habitats for the developing larvae and add oxygen into the water. Emergent and floating plants provide perching and egg-laying sites for the adults and emergence sites for the larvae. Trees, shrubs and ground-level plants near and around the ponds provide physical structures for perching, foraging, basking and roosting, and also help to shade the pond from the full force of the sun and excessive temperatures. As the dragonfly pond is part of an edible garden planted with different herbs and vegetables, there is an abundance of insect prey.

Around 20 species of fish have been introduced into the pond, including a few native to the Malay Peninsula such as the Pearl Gourami (*Trichopodus leerii*), Three Spot Gourami (*Trichopodus trichopterus*), Siamese Algae Eater (*Crossocheilus* species) and *Rasbora sumatrana*. Many waterbodies in Singapore have been plagued by non-native tilapia that were probably released by pet owners or fish-farm operators. Our small dragonfly pond is not spared. Tilapia are invasive and aggressive. A few times, I attempted to introduce small shrimps into the pond, but they were all eaten by the tilapia, which have eluded capture.

In a balanced eco-pond, aquatic plants, fish and microorganisms interact with each other and maintain a favourable environment. When sunlight shines on the pond, aquatic plants perform photosynthesis and produce oxygen and food for other creatures. Tiny aquatic insect larvae provide prey for carnivorous fish and dragonfly larvae. Frogs and toads prey on insects and other small creatures found on the floating and emergent plants, and also around the pond.

As the pond's ecosystem is working well, its original filtering system (which was needed when the pond contained only Japanese koi) was not necessary and has been removed. However, the pond's original water pump has been kept to generate a gentle water current, which many fish species enjoy. Maintenance required for the established pond is minimal. Once every two months or so, overgrown vegetation and algae are cleared and composted. Dirt on the bottom of the pond is siphoned out and used as fertilizer for the garden. Water changes are unnecessary. Water is replenished when there is an observable drop in the pond level due to evaporation. On an average day, four or five species of dragonfly can be seen in and around the pond. The number of species recorded so far (August 2021) is 29, including locally uncommon or rare species such as the Crenulated Spreadwing, Shorttail, Emperor, Sultan, Restless Demon, Bronze Flutterer and Sapphire Flutterer.

Watching & Photographing Dragonflies

Most dragonflies are sun-loving insects. Sunny days, from the late morning to mid-afternoon, are therefore ideal times for dragonfly watching. During these hours, many dragonflies, especially the conspicuous and colourful species that frequent open habitats such as ponds and reservoirs, can be easily spotted and observed with the naked eye as they feed, defend their territories or engage in reproductive behaviour. Binoculars can be used as an aid to observe the insects from a distance or help locate dragonflies that prefer to perch on the tops of trees (notably clubtails and many female skimmers), or hawk above the forest canopy (emeralds and cruisers).

Many of these dragonflies vanish from sight into the canopy or other retreats during overcast periods. Damselflies, which have a weaker flight compared to true dragonflies, are also known to be less active and to hide away during windy days, possibly to conserve energy. Dull weather can, however, offer a silver lining. Normally aerial true dragonflies such as the Wandering Glider, Banded Skimmer, Saddlebag Glider and Yellow-barred Flutterer often cease their foraging and descend to perch on low vegetation when the sky turns cloudy, thus allowing closer views. Moreover, if you are willing to wait until early evening, the diurnal assembly is replaced by a crepuscular cast, as duskhawkers, duskdarters and White-barred Duskhawks emerge from their daytime hideaways to patrol ponds, streams and swamps.

In closed habitats such as rainforests and swampy forests, dragonflies may be found at or close to streams, seepages or small pools, even temporary ones that vanish during dry spells, under the canopy. Many forest-dwelling species, such as the Common Flashwing, Handsome Grenadier and Dark-tipped Forest Skimmer, are also commonly encountered at sunlit trails or clearings, which provide basking or feeding spots, and will remain at their perches if you do not make sudden movements or brush against the vegetation. When alarmed, these dragonflies usually fly off and land on a higher branch nearby, but may return to the original perch after a while if they sense little or no disturbance.

There are also some species, such as the Malayan Grisette, Blue-spotted Flatwing, Shadowdancer and forest-dwelling duskhawkers, which require more effort to spot, as they prefer to lurk in dim and dense areas such as small gullies or masses of fallen branches. Such dragonflies can be more easily detected by signs of movement as the insects shift from one perch to another, possibly in response to an approaching shadow, than by painstakingly scanning every dark nook in the forest.

Photographing Dragonflies

Photography is a popular means of documenting these charismatic insects and sharing their delicate beauty. Smartphones have become, for many, the default choice for capturing and disseminating images. Coupled with a detachable close-up or 'macro' lens, they make it possible to obtain fairly high quality photographs of dragonflies, especially of species that tolerate close approach.

Many dragonflies, however, tend to flee when they sense an approaching human. Thus, some prefer to use a camera that can be coupled with lenses of longer focal lengths (100mm or more), which can yield frame-filling images at a greater and less intrusive distance from the subject. Such cameras include the traditional single lens reflex (SLR) models, as well as the newer and smaller mirrorless cameras that are likely to displace SLRs in the near future. There are also dragonfly watchers who opt for so-called bridge cameras, which offer a considerable range of focal lengths in a single package.

Marcus Ng

A rather obliging Dark-tipped Forest Skimmer, which allowed close approach with a smartphone equipped with a close-up lens.

The use of flashguns is a personal choice. Flashlight does not appear to harm or bother most insects, and may be useful to illuminate dragonflies that prefer darker corners of swamps and forests. A well-lit photograph, taken using a small aperture (f8 or more) that offers greater depth-of-field, can be useful in capturing anatomical details such as wing venation and the abdominal appendages. The drawbacks of flashguns include often harsh tones and unnatural-looking colours, which can be tempered by dialling down the flash output to a fraction of its full strength, and/or attaching a suitable diffuser (a piece of foam or other semi-opaque material that spreads out and tempers the light of the flashgun).

There are some who eschew flashguns altogether, relying on their camera's ability to capture detailed images under low-light conditions. Photographs taken without artificial lights may be grainier, but have the advantage of more natural colour tones – especially for dragonflies with highly metallic bodies such as flashwings and forest skimmers – and present the dragonfly as it is perceived by the naked eye. Tripods are seldom practical for photographing dragonflies given the insects' usually active habits.

A final comment on photography: many observers enjoy 'artistic' images with blurred backgrounds and a narrow point of focus, and there is plenty to admire in such photographs. However, for identification purposes, you should ideally capture sharp lateral (sidelong) views of the head, thorax and abdomen. Dorsal (top-down) shots, with the wings and their venation in focus, are also very helpful for distinguishing between very similar species of true dragonfly. For some groups of damselflies, such as the wisps (*Agriocnemis* species) and blue-coloured sprites (*Archibasis* and *Pseudagrion* species), detailed images of the abdominal appendages may be necessary to tell apart one species from another.

Marcus Ng

Male Common Flashwing, shot with natural light.

Marcus Ng

Male Common Flashwing, showing the effects of a flash on the wing colours.

Glossary

This glossary explains the key technical terms and local place names used in the book. It is adapted from Tang et al. (2010), with kind permission from the Lee Kong Chian Natural History Museum.

abdomen Hind section of an insect's body – long, thin and consisting of 10 segments in odonates. In odonates, **distal** segments bear primary reproductive organs and **anal appendages**, while second segment also bears male's **secondary genitalia**.
anal angle of wing Basal lower corner of hindwing. Usually acute (with a sharp angle) in males of the families Aeshnidae, Gomphidae and Corduliidae. Always rounded in both sexes of the Libellulidae.
anal appendages Also known as claspers or abdominal appendages. Appendages at end of abdomen used by male to clasp female during mating. Consist of pair of upper (superior)

appendages and either pair of lower (inferior) appendages in damselflies or single fused lower appendage in true dragonflies. Female odonates have just a pair of **cerci**, which are usually short except in the Aeshnidae and some Libellulidae. Cerci have a sensory function.
anal loop Group of cells surrounded by part of anal vein, located at basal rear section of hindwings of true dragonflies. Absent or rudimentary in species with very narrow hindwings. Well developed, and often shaped like a sock, in many members of the Libellulidae and Corduliidae.
andromorph Female dragonfly with colours that resemble male of her species. Used for species that are typically sexually dimorphic.
antenodal crossveins Crossveins in wing that run between costa and radius (the 1st and 3rd primary wing veins respectively), located between wing-base and nodus (*ante-* means 'before' in Greek). Important features used to distinguish certain dragonfly species are the number of antenodal crossveins or whether distal antenodal crossvein (the one closest to nodus) is complete (spanning two cells) or incomplete (spanning just one cell).
apical Situated at or towards tip. Tip of wing when applied to wing; section most remote from thorax when applied to abdominal segments. Opposite of **basal**.
auricles Small, ear-shaped projections on either side of second abdominal segment of males, and sometimes females, of true dragonflies of the families Aeshnidae, Corduliidae, Gomphidae and Macromiidae. Thought to play a role in guiding female abdomen during process of forming mating wheel.
basal Situated at or towards base. When applied to wings, refers to part nearest to thorax. When applied to abdominal segments, refers to parts nearest to thorax. Opposite of **apical**.
bukit Malay for 'hill'. Common element of local place names, such as Bukit Timah and Bukit Brown.
caudal Refers to features at tail or rear end, such as caudal gills of larval damselflies.
cerci Paired upper **anal appendages** that extend from 10th abdominal segment. Usually short and simple in females, but longer and often elaborate in male odonates.
clypeus Part of head below **frons** and above **labrum**. Greatly enlarged to form 'snout' in jewels (Chlorocyphidae).
compound eyes Large, round visual organs dominating head. The eyes are formed from hundreds or thousands of tiny facets called ommatidia. Each ommatidium is an integrated optical unit with its own cornea, lens and photoreceptor cells for distinguishing brightness and colour. Structure and size of eyes provide dragonflies with a wide field of view and good image resolution, which makes them particularly effective predators.
costa Primary vein that forms frontal or leading edge of wing. Important anatomical features along costa include the **nodus**, **pterostigma** and **antenodal crossveins**.
crepuscular Active at or around dusk, or around dawn.
cryptic Well camouflaged; blends well into background so as to prevent detection.
distal Away from base, outwards. See **apical**.
dorsal On or towards upperside. See **dorsum**.
dorsum Back or upperside of any part, especially synthorax and abdominal segments.
eclosion Emergence of adult insect from its pupa or (in the case of odonates) final larval stage. Also known as ecdysis.
endophytic Describes species that lay their eggs inside plant matter, notably damselflies and the true dragonfly family Aeshnidae.
entomologist Scientist who studies insects. Entomology is the scientific study of insects.
exophytic Describes species that drop or scatter their eggs directly into water or on to damp ground.
exuvia (pl. **exuviae**) Empty exoskeleton of final larval stage after adult has emerged.
femur (pl. **femora**) Third leg segment ('thigh') above **tibia** ('shin'), usually long and relatively thick.
frons Upper part of face ('forehead'), upwards facing in damselflies and angular in true dragonflies.
fugitive Term used to describe certain species that are rather elusive, irregular or unpredictable in occurrence.
hyaline Transparent; normally applied to wing membranes.
imago Adult (reproductive) form of an insect.

instar Phase between periods of moulting in development of insect larva.
jalan Malay for 'road' or 'street', for example Jalan Mashhor and Jalan Bahar.
kerangas Heath forest that grows on sandy, nutrient-poor soil.
labium 'Lower lip' of insects, visible on underside of head, covering mandibles ('jaws').
labrum 'Upper lip' of insects, located below **clypeus**.
lateral Side. Used to refer to features or colours on sides of thorax or abdomen.
lentic Term used to describe still waterbodies such as ponds, lakes and marshes. The opposite is 'lotic', which refers to habitats with moving water.
lorong Malay for 'lane', for example Lorong Halus.
mangroves Coastal or riverine forests dominated by certain broadleaved trees, such as *Rhizophora* and *Bruguiera* species, and palms like *Nypa fruticans,* which are adapted to intertidal and estuarine conditions. During low tides, upper or inland zones of mangroves are usually exposed, forming muddy flats with small pools and creeks that serve as habitat of certain **stenotopic** dragonfly species.
nodus Visible 'break' or angle along costa or leading edge of wing, which provides structural reinforcement as well as flexibility. See also **antenodal crossveins**.
obelisk position Posture adopted by some dragonflies (especially skimmers), in which abdomen is held erect, often at an almost vertical position. This helps a dragonfly to keep cool by reducing surface area of body exposed to the sun's rays.
ocellus Simple eye made up of single small convex lens. In Odonata, three of these lie between or in front of compound eyes.
ovipositor Structure, often blade-like, near tip of abdomen, used to insert eggs into plant tissue. Present in females of all damselflies and true dragonflies of the family Aeshnidae.
petiole Stalk. In odonates, usually refers to slender, stalk-like basal section of wings of many damselflies. **Anal appendages** may also be stalked or petiolated (as in the Leaftail).
phoresy Use of another organism for transport from one habitat to another. Phoretic organisms may be, but not necessarily, also parasites.
phytotelm (pl. **phytotelmata**) Water-filled plant cavities such as broken bamboo stems, tree holes and buttresses that serve as breeding sites for certain species, such as the Bombardier and Dryad.
postocular spots Contrastingly coloured, usually paired markings on back of head near or behind eyes.
pronotum Dorsal section of **prothorax**. May bear prominent lobes or 'spines' that are diagnostic in some species (for example *Agriocnemis* species).
prothorax Small, frontal portion of thorax, nearest to head, resembling a 'neck'. It bears forelegs, and in females of all damselflies and some true dragonflies, is grabbed by male **anal appendages** during mating.
pruinescence (pruinescent) Waxy-grey or bluish-white powdery secretion that develops on some parts of body and, rarely, wings, in some species when a dragonfly becomes mature; pruinescence may greatly alter or obscure a dragonfly's underlying colours.
pterostigma (pl. **pterostigmata**) Thickened, often expanded cell on costa or leading edge of wing near its tip. Often strongly coloured and stands out from other cells. The pterostigma acts as a counterweight, helping to regulate wing movements during flight.
pulau Malay term for 'island', for example Pulau Ubin and Pulau Semakau.
riffle Faster flowing, usually shallower, section of stream or river.
riparian Describes vegetation growing by sides of rivers and streams.
secondary genitalia Male sexual organs beneath 2nd abdominal segment, used for transfer of sperm to female. Just before mating, male first transfers sperm from his primary genitalia (below 9th abdominal segment) to the secondary genitalia. Female, after being clasped by male, curves her abdomen downwards so that her genitalia latch on to male's secondary genitalia.
seepage/seep Tiny pool or patches of groundwater within forests or at forest edges, often at base of a slope or overhang.
seta (pl. **setae**) Hair- or bristle-like structure.
stenotopic Describes species that can survive only in very specific habitats, such as mangroves

or shaded forest pools. The opposite is 'eurytopic', which describes species that can thrive in a broad range of environmental conditions and habitats.
subspecies Representatives of species inhabiting a particular geographical division of the range of the species and differing in appearance from those from other areas. Normally no two subspecies of the same species can occur in the same locality, as they would interbreed. Subspecies are given a trinomial taxonomic name, for example *Argiocnemis rubescens rubeola*, where *rubeola* is the subspecific name. An equivalent term is 'geographical race'.
Sundaland Biogeographical region of Southeast Asia that comprises the Malay Peninsula, Sumatra, Java, Borneo, Palawan and surrounding smaller islands. Eastern boundary of the Sundaland is Wallace's Line, which separates Borneo and Sulawesi. Distribution of several Singaporean dragonfly species is confined to Sundaland.
sungei/sungai Malay for 'river', for example Sungei Buloh (Buloh River) and Sungei Cina.
synonym In **taxonomy**, a scientific name now regarded as invalid, often because the species in question had been described earlier and the scientist who (later) coined the synonym was not aware of the earlier description. It could also be due to taxonomic revisions, whereby two species that were hitherto thought to be different were found to be conspecific, with the later described 'species' relegated to a synonym as a result.
synthorax Larger, main section of odonate thorax, formed by fused mesothorax and metathorax, bearing two pairs of wings above (dorsally) and midlegs and hindlegs beneath (ventrally). Dorsal and/or lateral marking of synthorax often diagnostic.
taxonomy Science of naming, describing and classifying living organisms, using system developed by Linnaeus whereby every animal or plant (also known as a taxon, pl. taxa) is assigned a binomial name consisting of a generic epithet (genus, pl. genera) and specific epithet.
terminal Situated at or towards tip, see **apical**.
teneral Newly emerged adult dragonfly, recognized by its soft, pale body and shiny wings, which has not yet developed colours of adult. Tenerals tend to move away from water to feed and mature. In the tropics, it probably takes from a few days to two weeks before an adult dragonfly is mature enough to breed.
tibia (pl. **tibiae**) Leg segment below femur, analogous to shin. Usually long and thin, and bears spines used in prey capture, which may be long and fine or short and robust.
type Specimen that was used by a scientist to describe and name a particular species. Singapore is the source of a number of odonate type specimens that were first collected by Wallace and later described and designated as types by scientists such as Sélys. The type locality is the place where the type specimen was collected.
venation Network of veins in wings. Some dragonflies have very dense venation with many crossveins and heavy reticulation, while others have open wing venation with relatively fewer crossveins.
ventral Refers to features on lower surface or underside of thorax or abdomen, such as ventral flaps and ventral spines.

Species Descriptions

The following section features all 136 dragonfly species known to occur in Singapore, including the 10 that have become locally extinct. It is organized by family, starting with those under Zygoptera, with a brief description of the general traits of each family, followed by individual species descriptions ordered alphabetically by the scientific name. Each species description provides the following information.

Scientific Name

This is the species' scientific name, following the taxonomic naming system devised by Swedish naturalist Carl Linnaeus, which provides every described species with a binomial name consisting of a generic epithet, the first letter of which is always capitalized, and a specific epithet, which is never capitalized. The binomial name is always italicized when printed or underlined when written.

The species authority, that is the person who first described and named the species, and the year this was done, is included to aid further research, for example *Podolestes orientalis* Selys, 1862. If given in parentheses, for instance *Neurobasis chinensis* (Linnaeus, 1758), this means that the species was originally described using a different generic epithet (*Libellula chinensis* in this case), and was later transferred to another genus.

Common Name

This is the locally accepted English name of the species. For consistency, this book follows – with minor modifications, for example 'adjudant' to 'adjutant' – the names used in the most recent Singapore checklist by Soh et al. (2019), which are in turn based on those established in the first field guide to Singapore's dragonflies by Tang et al. (2010).

Size

By convention, a dragonfly's dimensions include two aspects: the hindwing length (HWL) and total body length (TBL), inclusive of the anal appendages. This book takes reference from the male HWL and TBL measurements given in Tang et al. (2010), as well as measurements made using Robin Ngiam's specimen collection. Adult sizes may vary considerably, depending on the availability or quality of food during the larval stage. For some species, the term 'c.', short for circa (Latin for 'about'), is used.

Description

This gives the key physical characteristics of the species, including the eye and body colours, and distinguishing features of the body, wings and anal appendages, to aid identification in the field. The descriptions and accompanying images focus on mature adults of both sexes. However, where relevant and known, immatures and older adults are also described or depicted, as these may differ from typical adults, and some species, such as the Variable Sprite and Variable Wisp, undergo significant colour changes as they mature. Where available, images of older adults with blue-grey pruinescence that obscures their underlying colours are also included.

The description also provides information for telling apart very similar and potentially confusing species in the field, along with visual aids (line drawings or close-up photographs) that highlight key diagnostic features like the anal appendages. The descriptions are broadly applicable to each species as it occurs in Singapore, as well as the Malay Peninsula and Borneo. Subspecies or populations found beyond this region may differ quite considerably in colouration, for example subspecies of *Cratilla lineata* and *Rhyothemis phyllis*.

Habitat & Habits

This section describes the types of habitat preferred by the species, along with field observations of its habits, which may aid identification or simply be of interest.

Presence in Singapore

This section lists the known localities for the species in Singapore. The localities provided may not be exhaustive, especially for the more common and widespread species, as the aim is to furnish examples of suitable habitats. It should be noted that distribution records are highly subject to changes, as some species – even forest-dependent dragonflies – may extend their range into new areas that have become amenable to their needs. Conversely, some species may vanish from a particular site or fail to establish themselves after an initial appearance, due to changing or unsuitable environmental conditions. Dragonflies, including forest species, may also inadvertently turn up in unexpected urban areas such as high-rise buildings, perhaps drawn to lights at night. Such sightings should be classified as vagrants or accidental encounters rather than locality records.

Etymology

As a deeper layer of information, the meaning of a species' scientific name, where available or deducible, may be provided, along with a brief outline of its taxonomic history. Many elements of dragonfly names, such as *-themis* and *cora*, originate in European classical (Greek and Roman)

Marcus Ng

Rather old and worn male Common Parasol, with what appears to be algae growing on the abdomen and wings.

Marcus Ng

Young female Variable Sentinel. Mature females come in three forms (pp. 270–271).

Marcus Ng

Very old female Grenadier covered with pruinescence that obscures her underlying patterns.

Tang Hung Bun

Old female Common Bluetail with rather faded colours.

mythology, but other elements may provide descriptive indications of the colour, for example *rufa* and *glaucum*, or a significant anatomical characteristic (*Urothemis* and *capricornis*) and thus serve as mnemonic aids to recognize a species.

Distribution

This highlights the wider distribution of the species beyond Singapore, referenced using information from the IUCN.

National Conservation Status

This provides the species' conservation status in Singapore, based on Soh et al. (2019) and a recently completed new edition of the Singapore Red Data Book, an assessment of the country's threatened species.

IUCN Red List Status

The IUCN Red List status of the species is given to place the conservation of Singapore dragonflies in a wider global context.

Larva

A general description and image of the larva is included where available. Additionally, with kind permission from the Lee Kong Chian Natural History Museum, images of some species can be viewed via a Quick Response (QR) code to the museum's Biodiversity of Singapore (BOS) website and database. When scanned by a QR code reader (a feature of most smartphones), this leads to a BOS webpage that shows detailed and magnified images of the species' larva.

ZYGOPTERA

Argiolestidae (Flatwings)

This is a family of medium-sized to large damselflies with fairly long legs. The wings are stalked and hyaline, bear pterostigmata and are held very widely apart when perched, hence the common name 'flatwing'. The family consists of at least 116 species, mostly in Southeast Asia and Australasia, with a few representatives in Madagascar and Central Africa. The Southeast Asian genus *Podolestes* contains nine species, with four in the Malay Peninsula and five (four endemic) in Borneo.

The only local species in this group was previously classified under the family Megapodagrionidae (from *megapodagrion*, Latin for 'large-legged damselfly'). However, in 2013, the megapodagrionid subfamily Argiolestinae was separated to form its own family, Argiolestidae, based on the structure of the larval gills, which form a horizontal fan. This is a feature unique to the family; nearly all other damselflies have larvae with vertical gills.

The family name combines two terms of Greek origin. The prefix *argio-* may have been derived from Argia, a princess of Argos, an ancient Greek city, who became the daughter-in-law of Oedipus, king of Thebes, another Greek city. It could also have referred to Argia, a naiad (water nymph) in Greek mythology. The element *-lestes* (modified into the suffix -lestidae when applied to family names) is commonly applied to many damselflies that perch with their wings held open, such as the spreadwings (Lestidae) and flatwings (Argiolestidae). It means 'robber' in Greek and may refer to the insects' rapacious habits.

Marcus Ng

Male Blue-spotted Flashwing basking at the edge of a shaded swamp in the late afternoon.

BLUE-SPOTTED FLATWING *Podolestes orientalis*

Selys, 1862

Size Male HWL: 26–28mm; TBL: 42–46mm

Description Moderately large but lightly built damselfly with fairly long legs. Wings are hyaline and held wide open when perched, forming an 'X' when viewed from the front. Eyes light brown and blue, appearing greenish from some angles. Male's thorax has variegated light blue and slightly darker markings. Abdomen dark brown, with light blue banding at the basal end of each segment, and blue dorsal spot on segment 9; blue spot may be obscured with age. Female similarly patterned. Male's appendages prominent and darkish. Distal half of upper appendages sharply bent downwards. Lower appendages shorter, with prominent tuft of setae on underside. Young adults have all-blue eyes and white pterostigmata, which turn brown with maturity. Easily distinguished from local spreadwings (Lestidae), which also hold their wings wide open at rest, by horizontal perching posture; compare with the Crenulated Spreadwing (p. 132), which typically clings to vegetation with abdomen pointing downwards.

Habitat & Habits Forest-dependent species that breeds in well-shaded pools with rich leaf litter in swampy forests. May also occur at small, slow-flowing forest streams. Inconspicuous by virtue of its habitat as well as its habit of perching low and deep amid dense vegetation. Both sexes may occupy low twigs by or above small pools in swampy forests, from which they make sallies after prey. Flees into even darker corners when disturbed, but may return to a favourite perch after a while. Typically shuns direct sunlight, but both sexes may venture out to bask at edges of their microhabitat in the late afternoon. Courtship not apparent, with male seizing a willing female without any displays. After copulation, male disengages but stays close by, while female uses her ovipositor to probe and pierce a plant stem overhanging water, into which eggs are laid. She may also oviposit into plant matter such as dry leaves next to shallow forest pools.

Presence in Singapore Recorded in the Central Catchment Nature Reserve and adjacent nature parks such as Windsor Nature Park and Thomson Nature Park.

Etymology *Podo* is Latin for 'foot' or 'leg', and probably refers to the damselfly's long limbs. The suffix *-lestes* (Greek for 'robber') is a common element in the names of flatwing damselflies. The specific epithet is derived from *oriens*, Latin for 'sunrise' or 'east', and refers to the damselfly's origin in the oriental or far eastern regions (relative to Europe).

Distribution Singapore, Malay Peninsula, Borneo and Sumatra.

National Conservation Status Vulnerable; Restricted and Uncommon.

IUCN Red List Status Least Concern.

Larva Robustly built with long legs and three expanded caudal gills that form a horizontal fan. Hunts among fallen leaf packs at shallow edges of forest pools, with abdomen raised and gills aimed at the surface, possibly to maximize oxygen uptake in poorly oxygenated pools.

Tang Hung Bun

Close-up of the male's abdomen, showing the characteristic downwards curving upper appendages and a tuft of hair on the underside of the lower appendage.

Marcus Ng

A female, which can be told apart from the male by its thicker abdomen and prominent ovipositor.

Marcus Ng

Young male (note the white pterostigmata).

Marcus Ng

Young female.

Marcus Ng

Female inserting eggs into a dry leaf at the edge of a small swampy pool.

Robin Ngiam

The larva, showing its expanded caudal lamellae (left gill missing).

Calopterygidae (Demoiselles)

The demoiselles are large, slender damselflies with metallic bodies and very long legs. The wings are broad and unstalked, except in two genera in continental Asia, *Caliphaea* and *Noguchiphaea*, and may be brilliantly coloured or, if hyaline, exhibit iridescence under certain lighting conditions. Venation is usually very dense, with numerous antenodal crossveins. Pterostigmata are often absent, but females of some species have pale pseudopterostigmata (which have crossveins unlike true pterostigmata).

Worldwide, there are more than 180 species of demoiselle, with tropical and subtropical Asia having the greatest diversity. Demoiselles typically inhabit clear, unpolluted streams, and many species are dependent on forests. Their normal flight is bouncy and dance-like, with relatively slow, synchronized wingbeats, and usually brief, ending at the original perch or close by.

The family name, combining *kalos* (Greek for 'beautiful') and *pteryx* ('wing'), is a nod to the spectacularly patterned wings of many male demoiselles. Demoiselles, along with satinwings (Euphaeidae), jewels (Chlorocyphidae), grisettes (Devadattidae), flatwings (Argiolestidae) and several other damselfly families, are often placed in a superfamily called Calopterygoidea, which contains more than a quarter of all damselfly species.

Robin Ngiam

Calopteryx splendens (Harris, 1782), the Banded Demoiselle, is a common and fairly typical member of the family in temperate Eurasia.

White-faced Clearwing *Echo modesta* Laidlaw, 1902

Size Male HWL: 34–38mm; TBL: 56–60mm

Description Very large damselfly with strongly built, metallic greenish thorax and hyaline wings. Wings unstalked and bear dark pterostigmata. Eyes mostly dark brown. Mature males have prominent shiny white patch on frons. Females lack this white patch and have slightly darkened wing-tips.

Habitat & Habits Occurs at well-shaded streams in hilly rainforests. Where present, perches on streamside rocks or vegetation, but usually keeps well to the shadows, with the male's white frons the best giveaway to his presence. Males defend sunlit patches in their territories at the height of day.

Presence in Singapore Known locally from a single specimen collected between 1903 and 1914 by René Martin, a French entomologist. The specimen is now in the National Museum of Natural History in Paris, France, and was rediscovered by Matti Hämäläinen in 2012.

Etymology *Echo* is the name of a nymph (water fairy) in Greek mythology. Many odonates were named after such classical figures. However, it is now thought that the Belgian odonatologist Baron Michel Edmond de Sélys Longchamps coined the genus in memory of his daughter Marguerite, who died in 1852 aged four. Sélys had written 'Écho Marguerite' in his 1854 *Monographie des Caloptérygines*, which included a description of *Echo margarita*, a species from north-east India. In French, *écho* means 'in memory of'. *Modesta* means 'moderate' or 'modest' in Latin.

Distribution Singapore (formerly), Peninsular Malaysia, southern Thailand and southern Myanmar.

National Conservation Status Extinct.

IUCN Red List Status Least Concern.

Larva Demoiselle larvae are more elongated and leggier compared to larvae of most other damselfly families. The larva of this species is fairly typical of the family, though more compact than larvae of the genera *Neurobasis* and *Vestalis*. It has sprawling, dark-banded legs and long, pointed antennae that are bent outwards. The caudal gills have serrated edges.

Erland Refling Nielsen

Male photographed at Fraser's Hill in Malaysia, showing its boxy thorax. The wings bear pterostigmata, unlike those of other local demoiselles.

Oleg Kosterin

Female photographed on Phuket Island, Thailand, showing its darkened wing-tips.

Robin Ngiam

The shiny frons is the best giveaway to the male's presence. This individual was observed in Fraser's Hill, Malaysia.

Erland Refling Nielsen

Male in a dimly lit forest stream in Malaysia.

Matti Hämäläinen

Specimen (labelled as Climacobasis lugens, an invalid synonym) from Singapore, rediscovered in Paris by Matti Hämäläinen.

Marcus Ng

Late instar larva photographed in a hilly stream at Fraser's Hill, Malaysia, showing the serrated caudal gills.

Green Metalwing *Neurobasis chinensis* (Linnaeus, 1758)

Size Male HWL: 32–36mm; TBL: 56–60mm

Description Large, unmistakable damselfly with very long, slender legs and brilliantly coloured wings. Eyes dark brown and light green. Male has metallic green head and body. Forewings hyaline. Hindwings metallic green to blue-green above with dark wing-tips; closed wings present a bronzey to purplish sheen. Male lacks pterostigmata. Female has duller green body and hyaline wings tinted by light yellow-brown veins. Female's wings bear white pseudopterostigmata and nodal spots.

Habitat & Habits Occurs at clear, moderate to fast-flowing streams in a variety of forested habitats, from boulder-strewn upland streams to broad lowland creeks with a sandy or pebbly bottom. Tolerates disturbed environments, like brooks by villages and cultivation, but habitats should have ample riparian vegetation that provides submerged root masses for oviposition. Where present, males are conspicuous as they make regular patrol or display flights over the water, often using only their forewings to propel themselves and keeping their brilliant hindwings open and level. Males also perform a wing-flicking display while perched on a rock or branch, opening their wings briefly to 'flash' their iridescent upper colours; this probably serves as a warning to other males or a signal to entice potential mates. After courtship and copulation, male leads female to a suitable oviposition site and guards her while she crawls underwater to lay her eggs in root masses.

Presence in Singapore The last Singapore record was in 1970. The only known local site in 'recent' times was a stream in the Upper MacRitchie basin, which became silted up due to the construction of an expressway in the early 1970s, causing the extirpation of the population there.

Etymology First described by Carl Linnaeus, who named it *Libellula chinensis*. Linnaeus had based his description on a colour painting of a specimen said to be from China; he bestowed upon it the generic epithet he used for all dragonflies, *Libellula* (now limited to a group of true dragonflies found in North America and northern Eurasia), and named the species *chinensis* ('from China'). In 1853, Sélys reassigned the damselfly to a new genus, *Neurobasis*, for the numerous crossveins (from *neuron*, Greek for 'nerve') in the basal space (between the radius/media and cubitus, also known as the median space).

Distribution Singapore (formerly), mainland Southeast Asia, Sumatra, South Asia and southern China.

National Conservation Status Extinct.

IUCN Red List Status Least Concern.

Larva Slim and very elongated body with very long antennae, legs and caudal gills.

Linnaeus described the Green Metalwing based on this painting from George Edwards' A Natural History of Birds (1750), *which depicts a 'Golden bird of paradise'* (Sericulus aureus) *and a male Green Metalwing. Due to the open wings, Linnaeus mistook the insect for a true dragonfly.*

Marcus Ng

Male photographed in Chiang Mai, Thailand. Note the very long legs and the bronzy underside of the wings.

Cheong Loong Fah

Male flicking its wings on a rock in the middle of a stream. Photo taken in Endau-Rompin National Park, Malaysia.

Erland Refling Nielsen

Male performing a courtship display, with its hindwings held open and the tip of the abdomen turned upwards. Photo taken in Taman Negara, Malaysia.

Marcus Ng

Female photographed in Huu Lien, Vietnam. The wings bear pale pseudopterostigmata and nodal spots.

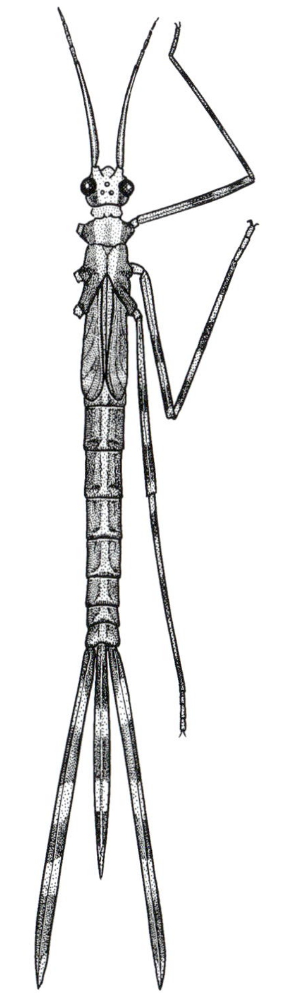
Albert G. Orr

Drawing of the elongated larva.

Common Flashwing *Vestalis amethystina* Lieftinck, 1965

Size Male HWL: 33–36mm; TBL: 50–56mm

Description Large, slender damselfly with metallic green body. Eyes black and green. Wings lack pterostigmata and are hyaline, but sparkle with purplish iridescence under certain lighting conditions and when exposed to direct flashlight. Sexes similar. Very similar to the Charming Flashwing (p. 64), but can be told apart by male's appendages; in the Common Flashwing underside of upper appendage bears small notch, while lower appendage is slightly hooked upwards. Female harder to distinguish; previously, the colour of the labium or lower lip (black in Common, yellow in Charming) was thought to be diagnostic, but recent studies suggest that this is not a reliable distinguishing characteristic. Best identified by association with males.

Habitat & Habits Found at clear, shaded forest streams and swamps with slow- to fast-flowing rivulets. Both sexes perch on vegetation by the water, making regular forays after prey and often returning to the same perch. Also commonly encountered basking and foraging along forest trails not far from streams. Towards midday, males take up positions at sunlit portions of a stream, awaiting females. Intruding rivals are challenged, with both males bobbing before each other in mid-air and circling upwards until one breaks off and the other claims or reclaims the territory.

Presence in Singapore Restricted to the Bukit Timah and Central Catchment Nature Reserves and adjacent nature parks. Quite easily encountered along trails in Windsor Nature Park and Dairy Farm Nature Park.

Etymology Genus is named after the Vestal Virgins, priestesses of Vesta, the Roman goddess of the hearth, home and family. Specific epithet (after amethyst, a violet form of quartz) may refer to iridescent quality of the wings, which arise from their structure rather than colour pigments.

Distribution Singapore, Peninsular Malaysia, Thailand and Sumatra.

National Conservation Status Vulnerable; Restricted but Common.

IUCN Red List Status Least Concern.

Larva Typical of family, but not as elongated as larvae of *Neurobasis* species, with fairly long legs and caudal gills, and outwards curving antennae. Black stripes and bands along thorax and legs. Each abdominal segment has two small black spots.

Marcus Ng

Male in a sun-lit portion of a forest trail.

Marcus Ng

The wings of both sexes sparkle with purplish iridescence when illuminated by a flashgun or certain angles of light.

Marcus Ng

Close-up of the male's anal appendages, showing the notch on the upper appendage and slightly hooked lower appendage.

Marcus Ng

A female, as evident by her abdominal ovipositor, foraging by a shaded forest trail.

Marcus Ng

Close-up of the male's head.

Robin Ngiam

The elongated larva of an unidentified flashwing.

CHARMING FLASHWING *Vestalis amoena*
Hagen in Selys, 1853

Size Male HWL: 32–35mm; TBL: 49–55mm

Description Large green demoiselle, very similar to the Common Flashwing (p. 62). Male can only be differentiated conclusively by the abdominal appendages. In Charming, the upper appendages lack a notch and the lower appendages are not hooked upwards. Female may be tentatively identified by association with males.

Habitat & Habits Found in forest streams and swampy forests. Describing the general behaviour of the genus, Orr (2003) wrote: 'Mating is preceded by courtship in which the male dances around the female displaying the sparkling colours of his wings.' He also observed (in Brunei), a female 'ovipositing unguarded in loose dead leaves, in a quiet rocky backwater among small stones'. Frederic Charles Fraser (1934) observed that 'unlike *Neurobasis*, the female [*Vestalis*] oviposits in blades of grass or juicy stems overhanging a stream, often several feet above the water's surface, the newly hatched larvae dropping from thence into the water'. It is possible that different methods of oviposition are used, depending on the site.

Presence in Singapore Both the Charming and the Common may occur together, but the former is less common locally and generally restricted to larger streams in Nee Soon Swamp Forest and the MacRitchie Reservoir area. Laidlaw (1902) remarked that '*Vestalis amoena* never occurs in the open, nor over rapidly running water', suggesting a preference for denser, swampier habitats.

Etymology Specific epithet comes from *amoenus*, Greek for 'lovely' or 'charming'.

Distribution Sundaland.

National Conservation Status Endangered; Restricted and Uncommon.

IUCN Red List Status Least Concern.

Larva Elongated and leggy, similar to larva of Common. Found among leaves and small stones at the edges of fast-flowing water.

Tang Hung Bun

Male in Nee Soon Swamp Forest.

Marcus Ng

Female flashwing, presumably amoena, basking at the fringes of Nee Soon Swamp Forest.

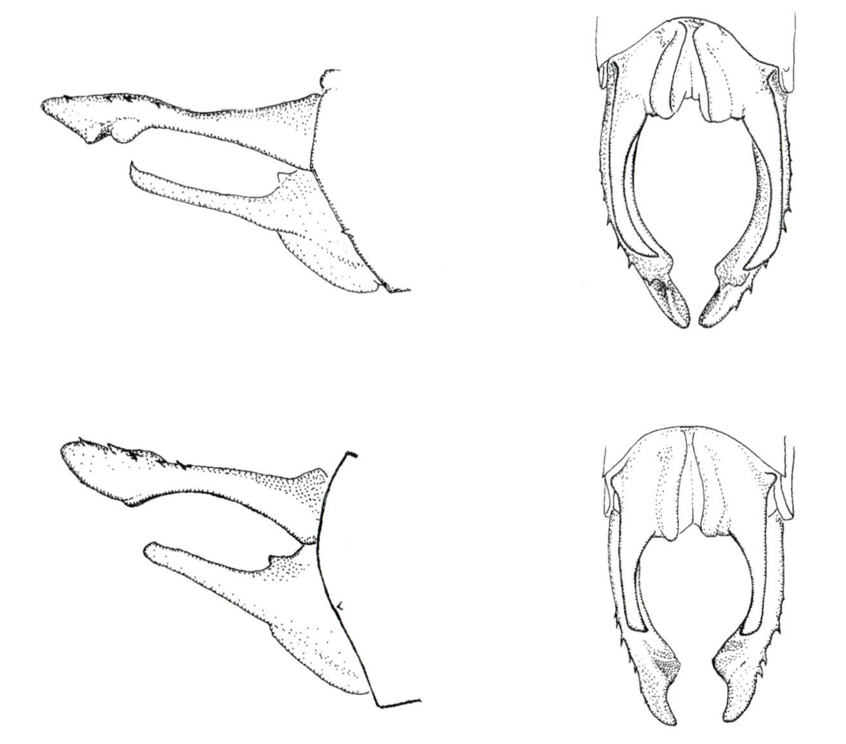

Albert G. Orr

Lateral and ventral views of male anal appendages of the Common Flashwing (upper row) and Charming Flashwing (lower row).

PLAIN FLASHWING *Vestalis gracilis* (Rambur, 1842)

Size HWL: 34–38mm; TBL: 60–63mm

Description Large green damselfly – largest of the three local flashwings – with hyaline wings that have iridescent highlights in certain lighting conditions, though less so than in the other two species. Eyes black and green. Thorax metallic light green, but green patches are clearly separated by thoracic sutures (lines that divide sections of synthorax). In the other two local flashwing species, thoracic green is largely contiguous.

Habitat & Habits Occurs at clear, shaded and sluggish forest streams, including somewhat disturbed habitats.

Presence in Singapore Elsewhere in region, fairly common in suitable habitats, but in Singapore, has been recorded at a single (and restricted) site in the northern part of the island, where it was first seen in 2012 and is locally abundant.

Etymology The specific epithet means 'slender' in Latin.

Distribution Mainland Southeast Asia and South Asia, as far north as Nepal.

National Conservation Status Critically Endangered; Restricted but Common.

IUCN Red List Status Least Concern.

Larva Similar in general appearance to larvae of the Common and Charming Flashwings (pp. 62 and 64).

Robin Ngiam

Male photographed in Singapore. Note the discontinuous metallic green patches on the side of the synthorax.

Dennis Farrell

Singapore forms the southernmost limit of the range of this flashwing, which is widespread in mainland tropical Asia. Photo taken in Thailand.

Dennis Farrell

Female photographed in Thailand. This species prefers well-shaded streams.

Chlorocyphidae (Jewels)

This is a unique and unusual family of small damselflies with abdomens that are shorter than the wings, except for males of *Rhinoneura*, a genus endemic to Borneo. The head also has a prominent upturned rostrum or 'snout' formed by the swollen clypeus. The wings are stalked and may be hyaline or partly coloured. There are numerous antenodal crossveins and pterostigmata are present.

To date, more than 160 species in 19 genera are known from tropical Asia, Africa and Australasia. Most jewels require clear, fast-flowing water in forested habitats. Many species have brilliantly coloured males, which perform dazzling display flights with rival males or during courtship.

Chloros means 'green' in Greek, while *cypha* (from *kyphos*) means 'curved' or 'hunchbacked', referring to the upturned 'snout'. *Chlorocypha*, from which the family name is derived, is a genus of jewels from Africa.

Robin Ngiam

Male in Nee Soon Swamp Forest, showing the brilliant abdominal colours.

Fiery Gem *Libellago aurantiaca* (Selys, 1859)

Size HWL: 15–17mm; TBL: 19–21mm

Description Small damselfly with fiery colours. Eyes blackish and dark brown. Male has black and yellow markings on thorax, and bright orange-red abdomen. Wings hyaline except for dark spot at forewing-tip, which appears iridescent blue under certain light. Anterior surface of femora and tibia (the main leg segments) has brilliant white pruinescence. Female has hyaline wings and yellowish or greyish-olive markings on thorax, matching male's.

Habitat & Habits Found at sluggish streams with sandy or silty bottoms, or quieter sections of swift, clear streams, in shaded forests. Usually perches close to the water on emergent twigs, half-submerged logs or floating leaves or sticks. Males establish small territories of 1.5–3m in diameter and engage in spectacular aerial displays, in which two or more males circle and make dashes towards each other, but without making physical contact. During such flights, forewings are often kept still and held at an angle to display their iridescent apical spots. This agonistic display is repeated until one male retreats. During courtship, male flies before a female with his abdomen arched upwards to display his colours, while extending and shaking his legs to show off the pruinescent surfaces. After copulation, female oviposits on partially submerged branches and leaves, while her mate remains close by to guard against other males.

Presence in Singapore Recorded in streams in Nee Soon Swamp Forest, where it can be quite common. There is an old record from Lower Peirce Reservoir in 1993, but it has not been seen there since. This species was described based on specimens collected in 1854 by Alfred Russel Wallace in Singapore (the type locality) and Johor.

Etymology *Libellago* is thought to combine *libellula* (Latin for 'dragonfly' in general) and *-ago*, a suffix derived from *virago* (Latin for 'heroine' or 'a woman with manly traits', *vir* being Latin for 'man'). Specific epithet comes from *aurantiacus*, Latin for 'orange' (the colour).

Distribution Sundaland and southern Thailand.

National Conservation Status Critically Endangered; Restricted but Common.

IUCN Red List Status Least Concern.

Larva Typical of genus. Body moderately elongated, though stouter than bodies of demoiselles, with dark banded legs and two long, thin caudal gills.

Tang Hung Bun

Two males performing an agonistic flight display.

Cheong Loong Fah

Female ovipositing on a twig in a stream.

Robin Ngiam

Larva of the Fiery Gem, which is fairly typical of the genus.

Clearwing Gem *Libellago hyalina* (Selys, 1859)

Size HWL: 16–18mm; TBL: 20–22mm

Description Small damselfly with subtle and variable, but still lovely colours. Slightly larger than other extant local gems, but rather less conspicuous due to more muted colours and lack of aerial displays. Eyes blackish and grey-green. Both sexes have hyaline wings. Male has dark thorax with thin yellowish markings on sides. Abdomen colours range from reddish-brown, bluish-grey to steel-blue and purplish, possibly depending on age. Legs black, contrasting with white leg segments of other local *Libellago* males. Female has pale thoracic markings and yellowish abdominal patterns that fade to grey with age; shape of irregular patch on side of thorax separates this species from other *Libellago* species.

Habitat & Habits Found at slow-flowing streams in shaded forests; also swamp forests. Males perch on low twigs or leaves over small pools and runnels, guarding small territories. During the hottest hours of the day, individuals may be seen perching high up in trees overlooking forest edges near streams. Territorial displays involve males hovering and rushing at their rivals but without making contact, and presenting their colours by wagging the abdomen up and down. Unlike in other local gems, there is no apparent courtship display; males simply seize and mate with females that enter their territories.

Presence in Singapore Locally restricted to Nee Soon Swamp Forest.

Etymology Specific epithet – from *hyalos*, Greek for 'glass' – refers to the fully hyaline wings, which are unusual for genus.

Distribution Sundaland (Sumatra, Borneo) and parts of mainland Southeast Asia as far north as Thailand and southern Vietnam, and possibly Myanmar.

National Conservation Status Critically Endangered; Restricted but Common.

IUCN Red List Status Least Concern.

Larva Typical of genus, found among leaf litter.

Robin Ngiam

Male with a violet abdomen.

Tang Hung Bun

Young female with a yellowish abdomen.

Tang Hung Bun

Male with a bluish-grey abdomen.

Erland Refling Nielsen

Male with dark colours photographed in Malaysia.

Tang Hung Bun

Male with a steel-blue abdomen.

Marcus Ng

Older female with darker colours.

GOLDEN GEM *Libellago lineata* (Burmeister, 1839)

Size HWL: 16–18mm; TBL: 19–21mm

Description Small, bright yellow damselfly with brilliant colours and flight displays. Eyes black and greyish-brown, paler in female. Male's thorax has alternating black and pale yellow markings; abdomen largely golden-yellow except for segments 7–10 and appendages, which are black. Legs black, but with brilliant white forwards-facing tibial surfaces. Wings hyaline except for dark forewing-tips, which appear iridescent blue under good light. Female has black and greyish-yellow markings. Two thin and parallel but incomplete yellowish lines on side of synthorax separate female from other local *Libellago* species.

Habitat & Habits Occurs at slow- to fast-flowing, open streams with sandy bottoms in forests; also reservoir edges. Adaptable to disturbed habitats such as canalized streams with flowing water and ample vegetation. Usually perched on low vegetation, rocks or half-submerged logs close to the water's surface, but may also be encountered at forest edges close to water, where both sexes forage for prey. On sunny days, males guard suitable oviposition sites and engage in prolonged non-contact aerial bouts similar to those of the Fiery Gem (p. 68). Two or more males fly in a semi-circular pattern around each other, keeping their forewings still and open at an angle, and moving backwards and forwards until one beats a retreat to a perch further away.

Presence in Singapore Formerly known from a single site in Mandai, but has since expanded its range widely, especially throughout the Central Catchment Nature Reserve. Recorded in various sites, including Upper Seletar Reservoir Park, MacRitchie Reservoir (Golf Link), Lower Peirce Reservoir Park, Windsor Nature Park and Bukit Brown (Jalan Mashhor).

Etymology Specific epithet, Latin for 'marked with a threadlike stroke', refers to thin, parallel lines on thorax.

Distribution Widespread in tropical Indo-Malayan region.

National Conservation Status Least Concern; Widespread and Common.

IUCN Red List Status Least Concern.

Larva Similar to other larvae in genus.

Marcus Ng

Male showing its prominent 'snout' and white leg surfaces.

Marcus Ng

The tips of the male's forewings turn iridescent blue in good light.

Marcus Ng

Female showing its characteristic thoracic markings, with two incomplete, overlapping lines.

Marcus Ng

Rival males performing an aerial display at a sunlit portion of a small forest stream.

Marcus Ng

Female ovipositing on a branch, with its mate standing guard close by.

ORANGE-FACED GEM *Libellago stigmatizans* (Selys, 1859)

Size HWL: 16–18mm; TBL: 20–22mm

Description Small damselfly with brilliant head markings. Eyes black and greyish. Male has bright orange between eyes. Thorax and abdomen black with apple-green markings on thorax and basal abdominal segments. Wings hyaline except for dark apical forewing-spot with blue iridescence. Tibiae have pruinescent white forward surface. Female lacks orange on head. Body blackish with pale greenish-yellow markings. Separated from the female Fiery Gem (p. 68) by lateral markings on abdomen, which are more triangular in shape.

Habitat & Habits Found in clear and swift forest streams with sandy or gravelly bottoms, especially around log jams (areas where fallen leaves, twigs and other plant debris accumulate), which provide perches and oviposition sites. Males engage in territorial flights similar to those of the Fiery and Golden Gems (pp. 68 and 72).

Presence in Singapore First collected in 1854 by Wallace in Singapore. Not recorded since.

Etymology Specific epithet comes from *stigmatizo*, Latin for 'to mark with a spot', and may refer to orange head marking.

Distribution Singapore (formerly), Peninsular Malaysia, southern Thailand and Sumatra.

National Conservation Status Extinct.

IUCN Red List Status Least Concern.

Larva Typical of genus.

Tang Hung Bun

Male photographed in Panti Forest Reserve in Johor, Malaysia.

Anthony Quek

Male hovering at a stream in Panti Forest Reserve, Malaysia.

Tang Hung Bun

Female ovipositing into a submerged log in Panti Forest Reserve, Malaysia.

Coenagrionidae (Pond Damselflies)
These are archtypical damselflies ranging from very tiny to fairly large species, of slender to medium build. Legs of local coenagrionids are proportionately shorter compared to those of other similar damselfly families such as the Platycnemididae, and bear short, sparse and heavy spines. The heads and bodies are often colourful. The wings are narrow, stalked and hyaline, with short pterostigmata. Sexual dimorphism (and also andromorphs) is quite common in the family.

The family contains more than 1,350 species worldwide, with 24 species recorded in Singapore. As their common name suggests, many live in ponds or still waterbodies in swampy forests or marshes, but some prefer flowing water. A few species, such as the Common Bluetail, Blue Sprite, Ornate Coraltail and Variable Wisp, are common in urban ponds and parks.

The family name comes from *Coenagrion*, which combines *coen-* (from *koinos,* Greek for 'common') and *-agrion*, a common element in damselfly genera. In Greek, *agrios* means 'wild' or 'living in the fields', and may reflect observations that these insects, unlike house flies, dwell far from human dwellings (at least in European contexts). The genus *Agrion* was originally coined in 1775 by Fabricius to denote all damselflies (Fabricius' teacher, Linnaeus, had earlier applied the genus *Libellula* to all dragonflies, including the damselflies). William Forsell Kirby then coined the genus *Coenagrion* and corresponding family name in 1890 to replace *Agrion*, which had become invalid for obscure taxonomic reasons.

Along with the featherlegs (Platycnemididae) and Isostictidae, a family restricted to Australasia, pond damselflies form a superfamily called Coenagrionoidea, which makes up about 60 per cent of all damselfly species.

Robin Ngiam

The Azure Damselfly, Coenagrion puella *Linnaeus, 1758, is a common pond damselfly found throughout Europe.*

BLUE SLIM *Aciagrion hisopa* (Selys, 1876)

Size HWL: 17–18mm; TBL: 31–32mm

Description Smallish blue damselfly with narrow wings and very slender build. Male has mostly blue eyes, darker on top, with blue postocular spots that are linked. Synthorax has black and blue stripes, with blue extending to basal segments of abdomen. Abdomen dark, except for segments 8–10, which are blue and may have black dorsal markings. Appendages black. Female similar but eyes greenish rather than blue. May be confused with blue *Pseudagrion* species, but the latter have totally blue eyes, usually with unlinked postocular spots, a heavier build and broader wings. *Aciagrion* contains more than 25 species in Asia, Africa and Australia, some of which are very similar looking. The genus is poorly understood taxonomically and in need of revision.

Habitat & Habits Occurs at shallow, weedy ponds, streams and drains in disturbed and open areas.

Presence in Singapore Recorded sporadically in the Bukit Timah and Central Catchment Nature Reserves, Normanton Park (Kent Ridge) and Pulau Tekong. There are also records from Pulau Ubin and Singapore Botanic Gardens in the 1930s. Possibly more widespread but overlooked.

Etymology Generic epithet combines *akís* (Greek for 'needle') and *-agrion*, a common suffix for damselfly names. It refers to the insect's exceedingly long and slender abdomen. Etymology of *hisopa* unclear: it could refer to the hyssop (*Hyssopus officinalis*), a herb with deep blue flowers. The naming of damselflies after flowers is not unknown; the Wandering Wisp was once named *Agriocnemis hyacinthus*, after the hyacinth.

Distribution South and Southeast Asia.

National Conservation Status Vulnerable; Restricted and Very Rare.

IUCN Red List Status Least Concern.

Larva Unknown but probably typical of genus, with rectangular-rhombus-shaped head and three distinctive leaf-shaped caudal gills.

Dennis Farrell

The Blue Slim has a markedly more slender build than the similar sized Blue Sprite. Photo taken in Thailand.

Cheong Loong Fah

Female from Pulau Tekong with somewhat muted colours.

Bennett Tan

Dead male (possibly killed by a spider) in MacRitchie Reservoir, showing its needle-like abdomen.

Dennis Farrell

Pair in wheel. Photo taken in Thailand.

Bill Ho (AFCD)

Larva of Aciagrion approximans*, a closely related species. Photo taken in Hong Kong.*

VARIABLE WISP *Agriocnemis femina* (Brauer, 1868)

Size HWL: 9–10mm; TBL: 20–22mm

Description Tiny damselfly with markings that vary greatly with age in both sexes. Eyes dark brown and light green. Thorax of young male has light green and black stripes. Abdomen mostly dark except for terminal segments and appendages, which are orange-red. Mature male develops powdery white pruinescence on synthorax and legs, and loses orange on abdomen and appendages. Distinguished from the very similar Wandering Wisp (p. 86) by lower abdominal appendages, which are markedly longer than uppers; in Wandering, male's upper appendages longer and have downwards pointing tip. Also separated by flattened posterior lobe of prothorax in male Variable. Young female cherry-red with dark dorsal stripe on thorax, but red turns to olive-green with maturity. Female also has an erect, bifurcated flap on posterior lobe of prothorax, visible when viewed laterally as tiny, erect 'spine'. Female Wandering has a much reduced posterior lobe that does not stick out.

Habitat & Habits Occurs at grassy edges of ponds, reservoirs, marshes, and shallow, boggy areas around slow-flowing streams and well-vegetated old ditches. Common and widespread damselfly that is probably overlooked due to its small size and habit of perching very close to the water. Tolerates disturbed areas and pools or drains with stagnant, somewhat polluted water, as long as there is ample fringing vegetation. Mature males, with their bright pruinescence, fairly easy to spot as they dart about like tiny floating pinheads. They remain active until around sunset. Mature females and young males less conspicuous but can usually be found close by, clinging to low vegetation.

Presence in Singapore Recorded in most green spaces in Singapore where suitable waterbodies exist, including the Singapore Botanic Gardens and urban parks such as Toa Payoh Town Park, Kent Ridge Park and Pasir Ris Park.

Etymology Specific epithet (Greek for 'female') refers to male's long lower abdominal appendages, which were thought to resemble an ovipositor. The Austrian entomologist Friedrich Moritz Brauer had originally named this species *Agrion femina*, using the generic epithet applied to all non-calopterygid damselflies then (see p. 76 for etymology of *Agrion*). Sélys later coined a new genus for this group of minute damselflies. He added the suffix *cnemis* (from *kneme*, Greek for 'leg'), because he thought that the genus was closely related to *Platycnemis*, a group of damselflies now placed in the family Platycnemididae.

Distribution Widespread in tropical and subtropical Asia and Australasia.

Marcus Ng

Mature male perched very low amid waterside vegetation.

National Conservation Status Least Concern; Widespread and Common.

IUCN Red List Status Least Concern.

Larva Typical of family, though very small, with three flattened caudal gills and occurring amid aquatic vegetation.

Marcus Ng

Young male with no hint of pruinosity. The long lower anal appendages are evident.

Marcus Ng

Male with developing pruinescence on the thorax and legs.

Marcus Ng

Mature females, with their greenish hues, are well camouflaged and harder to spot than males and young females.

Marcus Ng

Young female. The posterior lobe of the prothorax is visible as a tiny erect 'spine'.

Marcus Ng

Maturing female that is losing its red hues.

Marcus Ng

Older female with slight pruinescence on the thorax.

Tang Hung Bun

Copulation between a mature male and young female.

Tang Hung Bun

Young male and mature female in wheel.

MARSH WISP *Agriocnemis minima* Selys, 1877

Size HWL: 9–10mm; TBL: c. 22mm

Description Tiny damselfly, superficially similar to the Variable and Wandering Wisps (pp. 79 and 86), but easily separated by male's very long, downwards curving upper abdominal appendage, and green-centred black mark on dorsum of male's second abdominal segment. Eyes dark brown and light green. Male's synthorax has light green and black stripes. Pterostigmata light brownish-orange (greyish in Variable and Wandering). Abdomen-tip and appendages black; pale orange in younger individuals. Young female has orange-red thorax and abdomen. Mature female greenish with black markings; body develops blue-grey pruinescence with age.

Habitat & Habits Occurs in swampy forests and shallow marshes in open forests. Elsewhere in region, found at paddy fields, ditches and ponds.

Presence in Singapore First recorded locally in 2011 at an open marsh within the Central Catchment Nature Reserve, which remains its only known location.

Etymology Specific epithet means 'tiny' in Latin.

Distribution Southeast Asia.

National Conservation Status Critically Endangered; Restricted and Very Rare.

IUCN Red List Status Least Concern.

Larva General appearance typical of family, with three very obvious flattened, leaf-shaped caudal gills.

Dennis Farrell

Male in Thailand, showing the characteristic green dorsal spots on abdominal segment 2.

Dennis Farrell

Mature female is greenish with black markings. Photo taken in Thailand.

Dennis Farrell

Like others of the genus, younger females are reddish. Photo taken in Thailand.

Dennis Farrell

Male photographed in Thailand, showing its long upper appendages.

Cheong Loong Fah.

Male showing the long, downwards curving upper anal appendages.

Cheong Loong Fah.

Older females develop pruinescence over their bodies.

Dwarf Wisp *Agriocnemis nana* (Laidlaw, 1914)

Size HWL: 9–10mm; TBL: 19–20mm

Description Smallest damselfly in the region, and one of the smallest odonates in the world, dwarfed only by a few other closely related species such as *A. bumhilli* of southern Africa. Male has dark brown and light blue eyes. Synthorax light blue with greenish tinge dorsally, and with black stripes. Abdomen azure-blue with dark dorsal markings; segments 9–10 largely black. Appendages light blue. Pterostigmata bluish-white and grey. Female has dark brown and light green eyes. Synthorax light greenish-yellow with black stripes. Abdomen similar to male's. Pterostigmata greyish-brown. Younger females pale red.

Habitat & Habits Found around small streams, shallow, weedy ponds and grassy marshes in open areas close to forests. Known to occur in disturbed habitats. Minuscule damselfly that is possibly overlooked due to its inconspicuous size and habits. Perches very low amid dense waterside vegetation.

Presence in Singapore Found in the Central Catchment Nature Reserve, but also recorded in the Singapore Botanic Gardens in 2011.

Etymology Specific epithet comes from *nanus*, Latin for 'dwarf' or 'small'.

Distribution Mainland Southeast Asia.

National Conservation Status Endangered; Restricted and Very Rare.

IUCN Red List Status Least Concern.

Larva Unknown, but probably similar to that of other species in genus.

Marcus Ng

The sky-blue abdomen of the mature male, as well as its tiny size, readily distinguish the Dwarf Wisp from the other local wisps.

Marcus Ng

Late teneral male, which can be distinguished from the similarly sized Marsh Wisp (p. 82) by its stubbier abdominal appendages.

Marcus Ng

Post-teneral male with paler colours.

Marcus Ng

Like the male, the mature female Dwarf Wisp has a sky-blue abdomen.

Marcus Ng

Younger females are reddish.

WANDERING WISP *Agriocnemis pygmaea* (Rambur, 1842)

Size HWL: 10–11mm; TBL: 21–23mm

Description Tiny damselfly, very similar to and probably often confused with the Variable Wisp (p. 79). However, male's upper appendages longer than lower appendages, though not as long as those of the Marsh Wisp (p. 82), and also angled downwards at tip. This is the key diagnostic feature in the field for telling apart Wandering and Variable. Also, male Wandering has upright posterior lobe on prothorax, visible to the naked eye as tiny erect 'spine'. In male Variable, posterior lobe lies relatively flat against synthorax. In both species, older males develop whitish pruinescence on head and thorax. Female olive-green with black dorsal markings; young female pale red. Posterior lobe of female's prothorax much reduced, more like a flange, compared to erect, spine-like flap of female Variable.

Habitat & Habits Occurs on exposed grassy edges of ponds and marshes, both natural and man-made. Probably overlooked due to similarity with Variable and habit of perching low amid dense beds of reeds and grasses. Members of this genus, despite their tiny size and weak flight, are thought to undergo passive dispersal using air currents and trade winds that bring them to new habitats.

Presence in Singapore More common than previously thought, and recorded in various locations in recent years: Toa Payoh Town Park, Bishan-Ang Mo Kio Park, Punggol Barat, Pulau Semakau, Labrador Nature Reserve and MacRitchie Reservoir.

Etymology Specific epithet means 'small' in Greek.

Distribution Widespread from South Asia to Australasia.

National Conservation Status Least Concern; Widespread but Rare.

IUCN Red List Status Least Concern.

Larva Typical of genus.

Marcus Ng

Lateral view of a male, showing the upright rear lobe of the prothorax and downwards curving upper appendages.

Graham Reels

Mature female photographed in Hong Kong. Note the absence of an erect lobe on the prothorax.

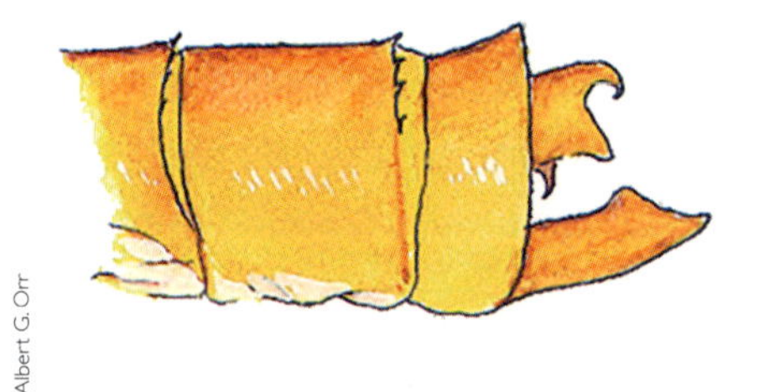

Albert G. Orr

Anal appendages of the male Variable Wisp, showing the much longer lower appendage.

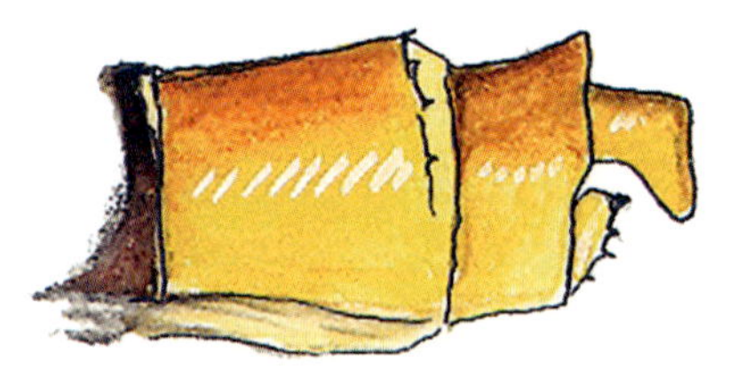

Albert G. Orr

Anal appendages of the male Wandering Wisp, showing the longer and hooked upper appendage.

Tom Kompier

Female *Variable Wisp* photographed in Vietnam, showing the erect, spine-like posterior lobe of the prothorax.

Tom Kompier

Female *Wandering Wisp* photographed in Vietnam. The posterior lobe of the prothorax lacks an erect 'spine'.

BEBAR WISP *Amphicnemis bebar*
Dow, Choong & Ng, 2010

Size HWL: c. 18mm; TBL: c. 38mm

Description Smallish, exceedingly slender damselfly, very similar to the more common Will-o-wisp (p. 90). Male dark metallic green and bronze-black. Abdominal appendages very pale and distinctive, with bifurcated upper appendages and very squat, barely discernible lower appendages. Young female red, turning bluish-green when mature. Distinguished from the Will-o-wisp by male's upper abdominal appendages, in which the upper arm lacks a small dorsal spine-like projection and has a pronounced downwards curve at the tip when viewed from side. Female differs by shape of posterior pronotal lobe (rear dorsal plate of prothorax): in the Bebar Wisp, lobe is relatively deeper, covering more of synthorax, with shorter dorsal spine.

Habitat & Habits Occurs in swampy forests with shaded streams and pools under a dense canopy.

Behaviour Very elusive damselfly due to inconspicuous appearance and habit of flying low fleetingly among swamp vegetation. At rest, usually perches at very tip of a leaf.

Presence in Singapore Locally, known only from Nee Soon Swamp Forest.

Etymology In Greek, *amphi-* means 'on both sides'. Sélys coined the genus for *A. wallacii*, a Bornean species, as he felt that it resembled *Amphilestes macrocephala* (now *Rhinagrion macrocephalum*), a damselfly with dark spots on both sides of the thorax. Suffix *-cnemis* a common element in damselfly names (see etymology of the *Agriocnemis*). Specific epithet refers to Sungai Bebar in Pahang, Malaysia, where the type specimens were collected in 2009. In 2011, five Singapore specimens collected by Murphy in 1994 as *Amphicnemis gracilis* were re-identified as belonging to this species.

Distribution Singapore and Peninsular Malaysia (Pahang).

National Conservation Status Critically Endangered; Restricted and Very Rare.

IUCN Red List Status Data Deficient.

Larva Unknown.

Cheong Loong Fah

Male in Nee Soon Swamp Forest, the only known location for the species in Singapore.

Robin Ngiam

Close-up of the male's anal appendages, showing the long and downwards curving upper arms, which also lack a dorsal spine.

Marcus Ng

Close-up of the male Will-o-wisp's anal appendages, which are straighter than those of the Bebar Wisp and have a small dorsal spine on the upper arm (absent in the Bebar Wisp).

Cheong Loong Fah

Close-up of the male's head and prothorax, showing an erect, spine-like lobe on the posterior lobe.

Rory Dow

Close-up of the female prothorax (head to the right), showing the posterior lobe, which extends further over the synthorax than in gracilis.

WILL-O-WISP *Amphicnemis gracilis* Krüger, 1898

Size HWL: 19–20mm; TBL: 40–42mm

Description Smallish damselfly with very light build and exceedingly slender abdomen. Narrow wings bear dark grey pterostigmata with paler edging. Male's eyes brown and light green. Synthorax dark metallic green on dorsum and upper flanks, pale yellow below. Abdomen dark, ending in very long, pale, almost translucent upper appendages subdivided into two thin arms. Upper arm expanded at tip, has small dorsal projection near base, lacking in the Bebar Wisp (p. 88), and does not curve strongly downwards at tip. Lower appendages stubby and barely discernible. Female's colours change greatly with age. Young female has brownish eyes and bright red thorax and legs, turning olive-green with maturity. Abdomen dark, except for pale segment 10 and cerci. Distinguished from Bebar by shallower posterior pronotal lobe and relatively longer dorsal spine. Young female may be confused with the Red-tailed Sprite (p. 122) from afar, or in dim swamps, where both species may occur together, but has a proportionately more slender profile. The Red-tailed Sprite also has a dark dorsal stripe on thorax and red abdominal tip.

Habitat & Habits Found in well-shaded, swampy forests. Like the Bebar Wisp, an inconspicuous damselfly that favours darker corners of swampy forests, small, sluggish streams, and vicinity of small, still pools filled with debris and fallen leaves. Both sexes may sometimes forage along shaded trails near swampy areas, resting on the tip of a leaf close to the ground. Seldom flies far when disturbed, but may be difficult to follow as it flutters in and out of shadows before settling again, usually close by.

Presence in Singapore Recorded in swampy portions of the Central Catchment and Bukit Timah Nature Reserves, and adjacent nature parks such as Windsor Nature Park and Thomson Nature Park.

Etymology Specific epithet means 'slender' in Latin. In European folklore, a will-o-wisp is a ghostly light seen in dark swamps and bogs.

Distribution Singapore, Peninsular Malaysia, southern Thailand and Sumatra.

National Conservation Status Vulnerable; Restricted but Common.

IUCN Red List Status Least Concern.

Larva Unknown.

Marcus Ng

Male is a mere sliver of an insect that seldom strays far from well-shaded swampy pools.

Marcus Ng

Close-up of the female's head and thorax, showing the rear pronotal lobe (and spine), which is shallower than that of the Bebar Wisp.

Marcus Ng

Female feeding on a moth.

Marcus Ng

Mature females are olive-green and highly inconspicuous.

Marcus Ng

Close-up of the male's head and thorax.

Marcus Ng

Despite their colours, young females can be inconspicuous as they perch very low in dim corners of swamps.

Blue-nosed Sprite *Archibasis melanocyana* (Selys, 1877)

Size HWL: 22–23mm; TBL: 37–38mm

Description Medium-sized, bright blue damselfly. Eyes black and bluish-green. Head and face with extensive light blue, hence the common name; large blue postocular spots. Male thorax has caerulean blue and black stripes. Abdomen blackish, but segments 8–10 bear caerulean blue markings; compare with Rebecca's Sprite (p. 94), which has mostly black segment 10. Appendages black. Sexes similar. *Archibasis* species may be confused with blue *Pseudagrion* species, but are easily told apart by their clearly bicoloured eyes, which are dark above and green or blue below; males of the Blue and Look-alike Sprites (pp. 115 and 113) have largely sky-blue eyes. *Archibasis* species are also markedly larger and more robustly built, with darker abdomens lacking blue banding and without extensive blue markings on segments 1–3. They also prefer well-forested habitats, whereas the other two species are found in more exposed, still waterbodies.

Habitat & Habits Occurs at small, fast-flowing streams in shaded swampy forests. Relatively weak flyer that tends to perch high (2–3m) above the ground. Very sensitive to movement from observers, reacting to disturbance by fluttering towards another high perch.

Presence in Singapore Recorded in swampy parts of the Central Catchment Nature Reserve. Also recorded from Windsor Nature Park. A single female was found at the National University of Singapore's Kent Ridge campus in 2012.

Etymology Generic epithet combines *arche* (Greek for 'beginning' or 'origin') and *basis* ('base'). Probably refers to wing-base, which has a narrow stalk. Sélys had originally coined the genus *Stenobasis* (*steno* means 'narrow') for the species in 1877, but it was later found that this epithet had already been used for a genus of true flies. Hence, Kirby coined *Archibasis* in 1890 to replace the invalidated genus. Specific epithet combines the Latin terms for 'black' (*melano-*) and 'dark blue' (*cyaneus*).

Distribution Singapore, the Malay Peninsula, Borneo and Sumatra.

National Conservation Status Endangered; Restricted and Rare.

IUCN Red List Status Least Concern.

Larva Unknown. General appearance should be typical of family and similar to larva of the Violet Sprite (p. 92).

Tang Hung Bun

Male showing extensive blue on abdominal segment 10.

Marcus Ng

Female in the swampy fringes of MacRitchie Reservoir.

Tang Hung Bun

Like the male, the female has extensive blue on the terminal segments of the abdomen.

Albert G. Orr

Drawing of the male anal appendages.

REBECCA'S SPRITE *Archibasis rebeccae* Kemp, 1989

Size HWL: 21–22mm; TBL: 35–37mm

Description Medium-sized blue damselfly. Eyes black and bluish-green. Male's synthorax has black and blue stripes. Abdomen black except for blue markings on segments 1, 2 (ventrally), 8 and 9. Segment 10 black, with two very small blue dorsal spots (may be absent in some individuals). Appendages black. Upper appendages strongly clubbed and hatchet-like when viewed from side. From above, each upper appendage expands to form inwards facing flange with flap-like extension. Female undescribed. Described in 1989 based on specimens from Pahang in Malaysia, but Lieftinck had collected a specimen (now in Naturalis Biodiversity Center, Leiden, the Netherlands) from Singapore or South Johor in the mid-twentieth century, which was earlier misidentified as a *Pseudagrion*.

Habitat & Habits Occurs at small streams with bottoms of fine sand and silt, in primary and secondary forests. Not recorded in swamp forests.

Presence in Singapore Recorded in the Central Catchment Nature Reserve.

Etymology Robert G. Kemp, who described the damselfly, named it after his daughter, Rebecca Louise Kemp, who accompanied him in the field.

Distribution Singapore, Peninsular Malaysia and Sumatra.

National Conservation Status Critically Endangered; Restricted and Very Rare.

IUCN Red List Status Near Threatened.

Larva Unknown. General appearance probably typical of family and similar to larva of the Violet Sprite (opposite).

Robin Ngiam

Marcus Ng

Close-up of the male's abdomen, showing the mostly black segment 10 and strongly clubbed upper appendages.

Male from Golf Link, MacRitchie Reservoir. Archibasis *species can be easily distinguished from blue* Pseudagrion *species by their heavier build and bicoloured eyes.*

VIOLET SPRITE *Archibasis viola*
Lieftinck, 1948

Size HWL: 23–25mm; TBL: 39–41mm

Description Medium-sized damselfly with unique violet-blue markings. Depending on the angle of light, colour can appear dark violet or blue, especially in photographs. Eyes black and blue; black and light green in female. Thorax has black and violet-blue stripes. Abdomen mostly black except for segments 8 and 9, which are violet-blue. Segment 10 and appendages black. Female similarly marked to male, but paler, more blue or greenish than violet.

Habitat & Habits Occurs at small, shaded forest streams and in swampy forests. Fairly common damselfly of slow-flowing streams, including somewhat disturbed habitats. Perches on low vegetation by the water. Female may also be encountered on vegetation along trails near streams.

Presence in Singapore Recorded in the Bukit Timah and Central Catchment Nature Reserves and adjacent nature parks; also other forested areas such as Bukit Brown, Jalan Bahar, Pulau Ubin and the Singapore Botanic Gardens (Marsh Garden).

Etymology Specific epithet means 'violet' in Latin.

Distribution Southeast Asia from Myanmar (possibly) to Palawan.

National Conservation Status Least Concern; Widespread and Common.

IUCN Red List Status Least Concern.

Larva Typical coenagrionoid in appearance. Dark brown body; caudal gills with leaf-vein-like patterns. Protuberant postocular lobes.

Marcus Ng

Male by a shaded forest stream.

Marcus Ng

Female Violet Sprites are quite variable in colour, with bluish to greenish hues on the thorax.

Marcus Ng

Male at a trickle in a swampy forest. The violet hues may show up as blue when a flash is used.

Marcus Ng

Female with greenish hues.

Marcus Ng

After mating, the male remains in tandem while the female inserts its eggs into aquatic vegetation.

Robin Ngiam

The dark brown larva of the Violet Sprite.

Variable Sprite *Argiocnemis rubescens rubeola* Selys, 1877

Size HWL: 16–18mm; TBL: 33–36mm

Description Small-medium damselfly with colours that vary with age in both sexes. Wings have rather pointed tips. Eyes black and light green. Mature male has light blue synthorax with black dorsal and lateral stripes. Abdomen light blue at base, grading to black. Segments 8–9 light blue; segment 10 and appendages black. Abdomen often slightly arched in profile. In younger males, blue on abdomen replaced by dull red, and thorax pale yellow to light green. Mature females similar to males, but abdomen-tip dark, except for tiny blue patch that is not always evident on side of segment 8. Younger females have red abdomen and black and yellowish markings on synthorax. Tenerals of both sexes have pale red abdomen and light yellow to light brown synthorax. May be confused with males of blue *Pseudagrion* species, but easily told apart by bicoloured eyes, thicker black stripes on synthorax and preference for well-shaded forested habitats. Distinguished from *Archibasis* species by smaller size, more slender build and usually slightly arched abdomen.

Habitat & Habits Found around ponds, marshes and drains with clear, stagnant water at forest edges; also slow, shaded streams in forest. Both sexes may be encountered in shaded clearings near streams.

Presence in Singapore Recorded in various locations such as the Central Catchment Nature Reserve, Windsor Nature Park, Ulu Sembawang Park Connector, Bishan-Ang Mo Kio Park and Pulau Ubin.

Etymology Prefix *argio-* probably an anagram of *agrio-* (from *Agrion*, Fabricius' catch-all term for damselflies), while *-cnemis* is a common suffix in pond damselfly names (see etymology of the Variable Wisp, p. 79). Specific epithet derived from *rubesco* ('to blush' in Latin) and means 'reddish'.

Distribution Widespread from north-east India to Australasia.

National Conservation Status Least Concern; Widespread but Uncommon.

IUCN Red List Status Least Concern.

Larva Unknown.

Marcus Ng

*Mature male can be distinguished from other blue-coloured sprites (*Archibasis *and* Pseudagrion *species) by the eye colours and slightly 'curved' abdomen.*

Marcus Ng

The reddish abdomens of young males are fairly conspicuous even in their shaded habitats.

Marcus Ng

Maturing male that has lost its red hues. Photo taken in Gopeng, Malaysia.

Tang Hung Bun

Mature female is similar to the male, except for the thicker and darker abdomen.

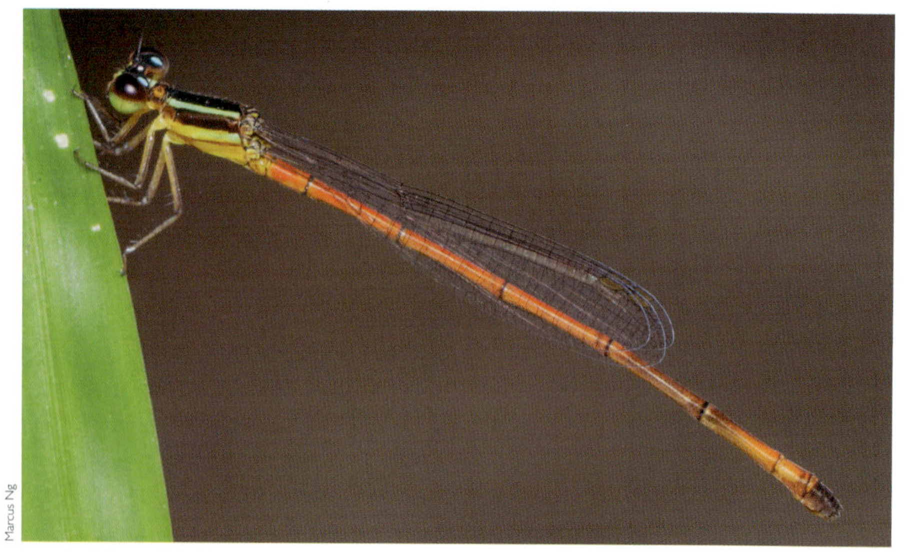

Marcus Ng

Young female has a reddish abdomen.

ORNATE CORALTAIL *Ceriagrion cerinorubellum* (Brauer, 1865)

Size HWL: 17–18mm; TBL: 35–38mm

Description Small-medium damselfly with unmistakable colours. Eyes bluish-green. Male's thorax greenish-blue. Abdomen mostly black except for orange-red on segments 1–2 and 7–10. Appendages very squat – a feature of the genus. Female similar but thorax more greenish than blue and abdominal colours paler.

Habitat & Habits Found in diverse well-vegetated habitats, including urban and rural ponds, marshes and open swamps within forests; also in semi-forested streams. Widespread and common damselfly, though less so than the Blue Sprite and Common Bluetail (pp. 115 and 103). Voracious predator of insects and even tenerals of other damselflies. Has been seen consuming smaller or even similarly sized damselflies such as *Archibasis* and *Pseudagrion* species.

Presence in Singapore Recorded in the nature reserves, adjacent nature parks, Pulau Ubin and many urban parks.

Etymology *Cerinum* is Latin for 'wax coloured' (pale yellow) and aptly describes many members of the genus. Here used in both the generic and specific epithets, combined respectively with *agrion* (a common element in damselfly names) and *rubellum* (Latin for 'reddish').

Distribution Widespread in tropical South and Southeast Asia.

National Conservation Status Least Concern; Widespread and Common.

IUCN Red List Status Least Concern.

Larva Light brown head and thorax, and dark brown abdomen. Caudal gills heavily patterned, with basal half shaded in light brown.

Marcus Ng

Male is fairly conspicuous among waterside vegetation.

Marcus Ng

Female is similar to male but has a more greenish thorax.

Marcus Ng

Female feeding on a grasshopper nymph.

Marcus Ng

Pair in wheel in a swampy forest.

Robin Ngiam

The dark brown larva of the Ornate Coraltail.

Fiery Coraltail *Ceriagrion chaoi* Schmidt, 1964

Size HWL: 16–17mm; TBL: 30–32mm

Description Brightly coloured damselfly, smaller than the Ornate Coraltail (p. 99). Male has red and green eyes. Thorax olive-green. Abdomen, including appendages, fiery red. Female has olive-green eyes and thorax, and dull green abdomen.

Habitat & Habits Found in ponds and slow-flowing streams with rich aquatic vegetation and dense surrounding vegetation.

Presence in Singapore Known only from a few localities, including MacRitchie Reservoir, Windsor Nature Park, the forest off Turf Club Avenue, Ulu Sembawang and a pond near Commonwealth Avenue West (Dover Forest). Also previously recorded in Bishan-Ang Mo Kio Park, but the population there appears to have been extirpated following a change in its pond-management regime.

Etymology Specific epithet honours Chao Hsiu-fu (1917–2001), the doyen of Chinese odonatology.

Distribution Mainland Southeast Asia.

National Conservation Status Vulnerable; Restricted and Rare.

IUCN Red List Status Least Concern.

Larva Similar to larva of the Ornate Coraltail, but caudal gills lack basal half shading.

Tang Hung Bun

Where present, the bright red abdomen of the male Fiery Coraltail is highly conspicuous.

Tang Hung Bun

Female is more drab than male.

Tang Hung Bun

Pair in wheel.

Tang Hung Bun

Pair in tandem at a pond in Bishan-Ang Mo Kio Park, where the species was formerly present.

Robin Ngiam

Tandem pair ovipositing in aquatic vegetation.

Robin Ngiam

The larva of the Fiery Coraltail.

COMMON BLUETAIL *Ischnura senegalensis* (Rambur, 1842)

Size HWL: 14–15mm; TBL: 28–30mm

Description Small but robustly built damselfly with polymorphic females. Male's eyes dark brown and green. Thorax light green, or bluish in some individuals, with black dorsal and lateral stripes. Abdominal segments 1–2 light green, with metallic blue-black patch on dorsum of segment 2. Segments 3–7 pale yellow, with dark dorsal line that expands to cover much of segment 7. Azure-blue 'tail-light' on segment 8 and sides of segment 9. Segment 10 and appendages black. Female has two colour forms. One has brown and green eyes and olive-green body; the other has brown-yellow eyes, with orange on thorax and base of abdomen. Both forms have dark stripe on dorsum of thorax and much of abdomen. Andromorphs fairly common, resembling males in every detail, including blue 'tail-light', except for lack of secondary genitalia on abdominal segment 2 and presence of ovipositor. Often found in the same areas as the Blue Sprite (p. 115), but can be told apart by its stouter, more compact build and thicker dark dorsal stripe.

Habitat & Habits Occurs in open and disturbed areas such as edges of ponds, reservoirs and drains, as well as wetlands with still or slow-flowing water. Less common in forested areas. May also be present near or in mangrove swamps. Both sexes can be found by the water's edge or nearby low vegetation, perching close to the ground. The blue 'tail-lights' of the male and andromorph females are very evident as they dart after prey or confront each other against a background of green or brown. Voracious predator that may capture other damselflies, including members of its own species.

Presence in Singapore Very widespread and common in urban and rural habitats, as well as grassy edges of forests.

Etymology Generic epithet combines *ischnos* (Greek for 'thin') and *oura* ('tail'). Despite this name, abdomen is relatively thick compared to abdomens of many other pond damselflies. Specific epithet, meaning 'from Senegal', indicates species' type locality.

Distribution Old World tropics from Africa to East Asia and Australasia. Has been recorded in Europe, which is outside its native range. Eggs or larvae were probably introduced by means of commercially distributed tropical aquarium plants.

National Conservation Status Least Concern; Widespread and Common.

IUCN Red List Status Least Concern.

Larva Typical coenagrionoid with variegated caudal gills.

Marcus Ng

A male. Note the compact build (compared to Pseudagrion species) and very short anal appendages.

Marcus Ng

Male with a more bluish thorax.

Marcus Ng

'Typical' olive-green female.

Marcus Ng

The somewhat less common but more conspicuous orange morph of the female.

Marcus Ng

Andromorph females can be recognized by the short ovipositor beneath the tip of the abdomen.

Marcus Ng

Mating wheel involving an andromorph female.

Marcus Ng

Mating wheel involving an orange female.

Blue Midget *Mortonagrion aborense* (Laidlaw, 1914)

Size HWL: c. 12 mm; TBL: c. 26 mm

Description Very small, slender damselfly, with colours that vary as the insect matures. Eyes black and green, with yellow postocular spots. Male thorax black, with thin greenish-yellow dorsal stripe (may be absent in some, probably younger, individuals) and two thicker greenish stripes on the side. Stripes are blue in older individuals. Abdomen dark with yellowish lateral streaks on segments 1–5 and faint banding thereafter. Segment 9 light blue, segment 10 with small blue marking. Upper appendages robust and club-like, longer than lowers. Female thorax dark (paler below), may lack dorsal stripe. Abdomen with yellow lateral streaks on segments 1–5 and thin blue banding at distal ends of segments 7–9 (no obvious 'tail-light'). Younger individuals of both sexes with blue postocular spots, paler eye and thoracic colours, and reddish abdomens. Currently it appears that there are two species of damselfly in the region that are being treated as the Blue Midget, for which the only consistent morphological difference between the two is in the structure of the penis. This requires further in-depth investigation. Moreover, the taxonomic relationship between the genera *Mortonagrion*, *Agriocnemis* and *Argiocnemis* requires a full review.

Habitat & Habits Shaded pools in swampy forest. Both sexes forage at the edges of sunny trails and clearings around mid-day, actively gleaning prey from leaves. Perches very low in undergrowth and fairly inconspicuous except for blue 'tail-light' of male and reddish abdomen of immatures. Retreats into dense vegetation or towards the canopy when disturbed. Mature female may perch very low at breeding sites (shaded forest pools).

Presence in Singapore First discovered in October 2021 by Shawn Ang. Currently known from a single site in the southern part of the island.

Etymology The specific epithet refers to the Abor region in north-east India, where the first specimen was collected.

Distribution North-east India, mainland Southeast Asia, Borneo and Sumatra.

National conservation status Critically Endangered; Restricted and Rare.

IUCN Red List status Least Concern .

Larva Unknown.

Marcus Ng

Male with greenish-yellow dorsal stripes on the thorax and blue abdominal 'tail-light'.

Marcus Ng

Mature female perched in the shadows.

Marcus Ng

Close-up of the male's anal appendages.

Marcus Ng

Male without the dorsal thoracic stripe.

Marcus Ng

Young male with a reddish abdomen.

Marcus Ng

Young female.

ARTHUR'S MIDGET *Mortonagrion arthuri*

Fraser, 1942

Size HWL: 14–15mm; TBL: 28–31mm

Description Small damselfly that can be distinguished from other similarly sized damselflies by very light build, distinctive postocular markings and habitat preference. Male has dark brown eyes that are paler below, with two conjoined bluish-grey marks behind each eye. Synthorax has black and greenish-yellow stripes. Abdomen dark brown with some pale banding and blue 'tail-light' on part of segment 8. Appendages black. Lower appendages much longer than uppers. Younger males have pale blue thoracic markings and were once thought to be a 'blue form'. Female has dark brown and pale green eyes. Postocular markings brownish-cream, larger and less well defined than in male. Thorax has pale brown and bluish-grey stripes. Some individuals have blue 'tail-light' on abdominal segment 8, but this is not evident in all individuals (possibly obscured by age or mud).

Habitat & Habits Found in mangrove forests and creeks. Highly stenotopic species restricted to sheltered pools and rivulets in the landward zones of mangroves. May also occur in freshwater pools near back mangroves. Inconspicuous due to slim build and muted colours, but adults can be found perched on pneumatophores (aerial breathing roots) of mangrove trees that rise from the mud. Males, which are more visible due to their blue 'tail-lights', guard tiny muddy pools that are subject to tidal influence.

Presence in Singapore Recorded in Admiralty Park, Loyang Mangroves (1987 record), Pulau Semakau, Pulau Tekong, Pulau Pawai and Pulau Ubin. Elsewhere in region, also found at streams in coastal forests that empty into the sea.

Etymology Genus named after Kenneth J. Morton (1858–1940), a Scottish entomologist. This species was named after Arthur Wheeler, a four-year-old who collected the type specimen in the garden of a seaside house in Butterworth (mainland of Penang state in Malaysia).

Distribution Singapore, Peninsular Malaysia, southern Thailand (Similan Island), Myanmar (Mergui Islands), and Sumatra (Belitung Island). Known from only about a dozen localities.

Marcus Ng

Mature male showing the yellowish thoracic stripes and blue 'tail-light' on the abdomen.

National Conservation Status Vulnerable; Restricted and Rare.

IUCN Red List Status Near Threatened.

Larva Distinguished from other local coenagrionoid larvae by less expanded (lateral view), sword-shaped caudal gills with 'V'-shaped mark.

Tang Hung Bun.

Close-up of the male's head, showing the distinct postocular markings.

Tang Hung Bun

Younger males are blue. Earlier, this was thought to be a different colour morph.

Marcus Ng

Youngish female perching on a mangrove aerial root.

Marcus Ng

Female, probably an older individual, that lacks the abdominal 'tail-light'.

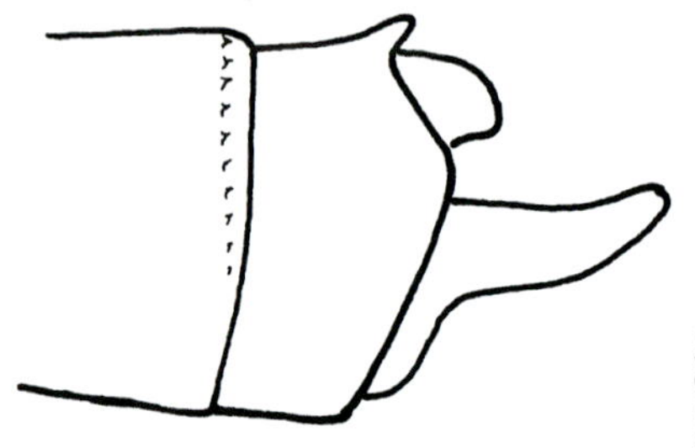

Albert G. Orr

Drawing of the male's anal appendages, showing the longer lower anal appendages.

HOOKED MIDGET *Mortonagrion falcatum* Lieftinck, 1934

Size HWL: 10–12mm; TBL: c. 21mm

Description Tiny blue damselfly with very delicate build. Eyes dark brown and green. Male has black and greenish-blue stripes on synthorax. Abdomen brownish, with blue 'tail-light' on part of segment 8. Appendages dark, with upper appendage forming strongly downwards curving hook. Female similar but blue on thorax less defined and lacks abdominal 'tail-light'.

Habitat & Habits Found in marshy edges of swamps and forests. Very rare and highly threatened species locally. A site in Tuas where it formerly occurred has been cleared for development, but it still survives in marshy portions of the central nature reserves. Inhabits shady areas of marshes with dense undergrowth, perching very low.

Presence in Singapore Recorded in the Central Catchment Nature Reserve and Tuas (formerly).

Etymology Specific epithet means 'hooked' in Latin and refers to shape of upper abdominal appendage.

Distribution Southeast Asia.

National Conservation Status Critically Endangered; Restricted and Very Rare.

IUCN Red List Status Least Concern.

Larva Unknown.

Cheong Loong Fah

Male of this very small and delicately built damselfly.

Cheong Loong Fah

The female, which is slightly more robust than the male.

Tang Hung Bun

Close-up of the male abdomen, showing the strongly hooked upper appendages.

DRYAD *Pericnemis stictica*
Hagen in Selys, 1863

Size HWL: 33–38mm; TBL: 62–66mm

Description Very large but slender damselfly. Eyes light green to brown-yellow. Dorsum of synthorax metallic bronzey-green, sides greenish-yellow. Abdomen brownish, with pale appendages. Sexes similar. Pterostigmata whitish-yellow with well-defined borders.

Habitat & Habits Occurs in forests with ample phytotelmata (water-filled plant bodies). Breeds in water-filled tree holes, leaf axils, buttress pans and bamboo stumps. Hence not restricted to streams and swamps, and may sometimes be seen along trails, especially in hilly areas with large bamboo groves and mature forest trees that provide suitable breeding sites. Largest extant damselfly in Singapore (genus includes largest pond damselflies in region), with habits that recall the giant helicopter damselflies (Coenagrionidae: Pseudostigmatinae) of South America. When active, floats about like a miniature helicopter, inspecting tree trunks and thickets for small prey. At rest, perches in a shady spot at tip of a leaf, with abdomen hanging downwards.

Presence in Singapore Recorded in several locations in the Bukit Timah and Central Catchment Nature Reserves, Dairy Farm Nature Park, Windsor Nature Park and Thomson Nature Park.

Etymology In Greek, *peri* means 'all around' and refers to the pterostigma, which has a well-defined border. The suffix *-cnemis* was used by Sélys as he originally thought that the genus was allied to *Platycnemis* (see etymology of *Agriocnemis*). Specific epithet means 'marked' and may also refer to the well-marked pterostigma.

Distribution Southern Thailand to Sumatra, Java and Borneo.

National Conservation Status Vulnerable; Restricted and Rare.

IUCN Red List Status Least Concern.

Larva Very dark and found among detritus in phytotelmata. Caudal gills unmistakable, being well expanded and broadly elliptical in shape. Larva regularly claps these gills rhythmically. Purpose of this not fully understood, but most likely to aid respiration in a low oxygen environment.

Marcus Ng

Male found in a small clearing by Wallace Trail near large clumps of bamboo.

Marcus Ng

Female in typical perching position at the tip of a leaf.

Robin Ngiam

Dorsal view of the larva.

Robin Ngiam

Lateral view of the larva, showing the distinctive elliptical caudal gills.

Look-alike Sprite *Pseudagrion australasiae* Selys, 1876

Size HWL: 19–21mm; TBL: 36–40mm

Description Small-medium blue damselfly with moderate build. Male's eyes sky-blue. Thorax light blue with thin black stripes on dorsum and upper side. Abdomen largely blue with dark dorsal markings and banding. Segments 8–9 light blue, with very narrow black band at distal end of segment 8. Segment 10 blue with small black dorsal mark. Upper appendages black, lowers pale; both appendages of nearly equal length and shorter than segment 10. Easily confused with the more abundant Blue Sprite (p. 115), but markedly larger and more robust. Also much thinner black band at end of segment 8, reduced black on segment 10, and shorter upper appendages. Female has greenish eyes. Thorax pale green with moderately thin dark dorsal stripe and brownish stripe that may be faint on side. Abdomen pale green with dark dorsal stripe. Distinguished from female Blue Sprite by greater size and thin dark dorsal stripe on thorax.

Habitat & Habits Occurs in exposed, grassy areas around reservoirs and open wetlands. Both sexes may be found perched amid reeds by the water or vegetation along forest edges.

Presence in Singapore Recorded in marshy portions of MacRitchie Reservoir, Windsor Nature Park, Bishan-Ang Mo Kio Park, Bukit Brown and Labrador Nature Reserve.

Etymology In Greek, *pseudo-* means 'false'. Generic epithet coined by Sélys, who found it hard to distinguish this genus from other damselflies then placed in the now-invalid genus *Agrion*.

Distribution South and Southeast Asia (Myanmar to Sumatra and Java). Also Hainan Island. Does not reach Australasia, despite specific epithet.

National Conservation Status Vulnerable; Restricted and Uncommon.

IUCN Red List Status Least Concern.

Larva Caudal gills have distinctive black spot at their middle portion and pale, snowflake-like markings along their margin.

Marcus Ng

In the field, the male Look-alike Sprite comes across as a slightly larger, more strongly built insect than the very similar Blue Sprite.

Marcus Ng

Dorsolateral view of a female Look-alike Sprite, showing the thin dark dorsal stripe, which is absent in the female Blue Sprite.

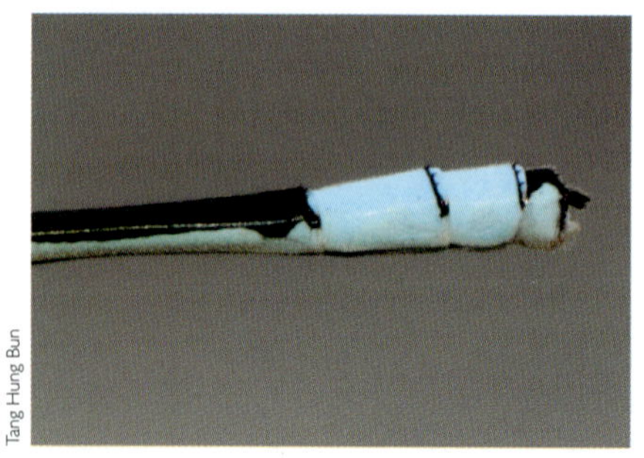

Tang Hung Bun

Abdomen of the male Look-alike Sprite. Note the short upper appendages.

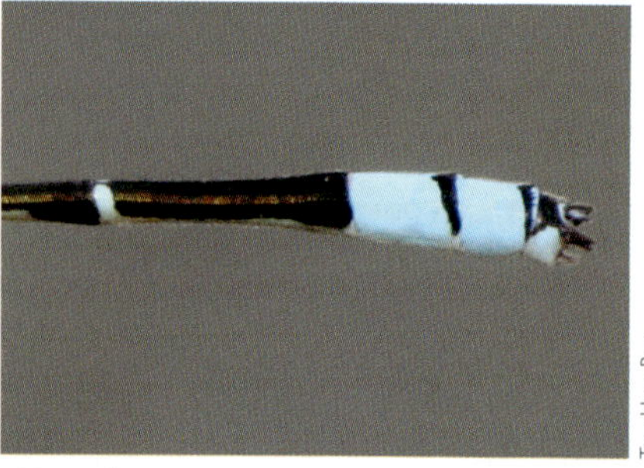

Tang Hung Bun

Abdomen of the male Blue Sprite, showing the greater amount of black and longer upper appendages.

Marcus Ng

Lateral view of female, showing its rather robust thorax.

Robin Ngiam

Green-coloured larva of the Look-alike Sprite.

BLUE SPRITE *Pseudagrion microcephalum* (Rambur, 1842)

Size HWL: 16–18mm; TBL: 30–34mm

Description Smallish blue damselfly with moderately light build. Male very similar to the Look-alike Sprite (p. 113), but markedly smaller and lighter in build. Also distinguished from Look-alike by upper abdominal appendages, which are longer than lowers and as long as segment 10. Black band on segment 8 also thicker, and segment 10 has a greater amount of black. Female has light brown and green eyes. Thorax pale yellow or greenish, with pale brown and light blue dorsal stripes. Abdomen greenish-blue with dark dorsal stripe. Distinguished from female Look-alike by smaller size, lighter build and lack of dark dorsal stripe on thorax.

Habitat & Habits Found in open habitats with still and slow-flowing water, including exposed edges of ponds, drains, canals, marshes and reservoirs. Tolerates disturbed and artificial habitats such as concrete-lined urban ponds with little vegetation. May also be found within forested areas close to open waterbodies. Possibly the most abundant local damselfly, along with the Common Bluetail (p. 103). Adults active by the water till early evening, with pairs staying in wheel for an hour or more. Female oviposits with male in tandem, crawling underwater to insert her eggs into aquatic vegetation. Immatures frequently encountered at forest edges and vegetation near ponds and reservoirs, and may stray into gardens and wayside planting. Active forager, gleaning small prey from foliage or hovering before spiders' webs to ram its legs at ensnared insects and possibly the web builder.

Presence in Singapore Found across the island in many urban parks and reservoir edges, as well as in the open and marshy portions of nature reserves and nature parks.

Etymology Specific epithet means 'small headed'.

Distribution Tropical Asia and Australasia from India to the Solomon Islands.

National Conservation Status Least Concern; Widespread and Common.

IUCN Red List Status Least Concern.

Larva Caudal gills largely shaded with light brown from base to mid-length, after which brown patterning is reduced from margins. Inhabits aquatic floating vegetation near the water's surface.

Marcus Ng

Mature male. The longish upper anal appendage (compared to the Look-alike Sprite's) is evident.

Albert G. Orr

Drawing of the male Blue Sprite's anal appendages.

Albert G. Orr

Drawing of the male Look-alike Sprite's anal appendages.

Marcus Ng

The female Blue Sprite has light blue and brown thoracic markings. It lacks the dark dorsal stripe of the female Look-alike Sprite.

Marcus Ng

Young males have light brown markings on the thorax and turn blue with age.

Marcus Ng

Pairs in wheel may stay together for an hour or more, flying deeper into the vegetation when disturbed.

Marcus Ng

A wheel involving a female with a blue-greenish thorax.

Marcus Ng

A female, in tandem with her mate, ovipositing into Hydrilla, *a common aquatic weed.*

Grey Sprite *Pseudagrion pruinosum* (Burmeister, 1839)

Size HWL: 21–23mm; TBL: 40–44mm

Description Medium-sized damselfly with unique colours. Pterostigmata reddish-brown. Male has reddish-brown eyes and mouthparts. Thorax and abdominal segments 8–10 covered with blue-grey pruinescence. Female has olive-green eyes and obscure olive-green and brown markings on thorax. Markedly larger than females of other local *Pseudagrion* species.

Habitat & Habits Found at fast-flowing streams and slower channels with grassy banks in open areas near forests. Stream-dependent, open-country species that is fairly common elsewhere in the region but less so in Singapore, where such rural habitats are now scarce. Where present, males perch very close to the water and are evident by their pruinosed colour as they dart about after prey or rivals. After copulation, pairs in tandem choose a perch above the water and slowly crawl backwards to a depth of about 10cm. Male often remains in tandem as female oviposits in aquatic vegetation.

Presence in Singapore Recorded in Mandai, Bukit Brown, Nee Soon Swamp Forest, Tuas and a few locations in the Central Catchment Nature Reserve.

Etymology Specific epithet means 'covered with hoar-frost' in Latin, and probably refers to blue-grey pruinescence of mature male.

Distribution Southern China to Southeast Asia, but not Borneo.

National Conservation Status Vulnerable; Restricted but Common.

IUCN Red List Status Least Concern.

Larva Colour varies from light brown to green. Can be distinguished from the Look-alike and Blue Sprites (pp. 113 and 115) by light brownish, blotchy caudal gills.

Marcus Ng

Male in Jalan Mashhor showing its red eyes and mouthparts.

Marcus Ng

Male showing its red eyes and mouthparts.

Tang Hung Bun

Grey Sprites in wheel.

Marcus Ng

Female photographed in Thailand.

Tang Hung Bun

Fully submerged female inserting its eggs into an underwater stem.

Bill Ho (AFCD)

Larva of the Grey Sprite. Photo taken in Hong Kong.

ORANGE-FACED SPRITE *Pseudagrion rubriceps* Selys, 1876

Size HWL: 18–20mm; TBL: 33–37mm

Description Small-medium damselfly with unique head markings. Male has bright orange face and partly orange eyes. Thorax light blue with olive-green on dorsum. Abdomen olive-green to light blue with dark dorsal line; segments 8–10 mostly light blue. Upper appendages black, lowers blue. Female has greenish eyes. Thorax light green or orange-yellow (paler on lower half), with faint brownish stripes on dorsum. Abdomen light green with dark dorsal line; abdominal segments 9–10 have blue dorsal markings.

Habitat & Habits Found at ponds and slow-flowing streams with ample marginal vegetation. Elsewhere in region, found in open and disturbed habitats, including brackish channels on landward sides of mangroves. Where present, males are evident by their bright orange faces, as they fly low over the water. Females much less conspicuous and more easily seen as part of a mating pair. After copulation, female partially submerges herself to oviposit in submerged root masses with male in tandem.

Presence in Singapore Known localities have greatly increased in the past decade. Recorded in Nee Soon Swamp Forest, Bukit Brown, Toa Payoh Town Garden, Punggol Waterway, Lorong Halus and HortPark.

Etymology Specific epithet means 'red-headed'.

Distribution South and Southeast Asia.

National Conservation Status Near Threatened; Widespread but Rare.

IUCN Red List Status Least Concern.

Larva Latter half of caudal gills features blotchy patterns.

Marcus Ng

The male is conspicuous and easily recognized by its brightly coloured head.

Robin Ngiam

A female.

Marcus Ng

Orange-faced Sprites in wheel. Photo taken in Gopeng, Malaysia.

Bill Ho (AFCD)

Dorsal view of the larva. Photo taken in Hong Kong.

Bill Ho (AFCD)

Close-up of the larva's blotchy caudal gills. Photo taken in Hong Kong.

CRYPTIC SHADESPRITE *Teinobasis cryptica* Dow, 2010

Size HWL: c. 20mm; TBL: c. 35mm

Description Very slender and delicate damselfly with emerald green eyes and greenish body. Male's synthorax pale bluish-green with faint brownish markings. Abdomen very slender and mostly pale greyish-blue-green on sides. Segments 9–10 black. Upper appendages black, lowers brown with black tip. Female's synthorax green-brown, paler below. Abdomen brownish with black terminal segments. Synthorax of both sexes has pale dorsum, distinguishing this species from other *Teinobasis* species in region.

Habitat & Habits Occurs in swampy forests. As its name suggests, a very cryptic and inconspicuous species that perches low in the shade of understorey vegetation. Both sexes hang from undersides of leaves. Females have been seen ovipositing into vegetation just above the surface of swampy pools.

Presence in Singapore Recorded in Nee Soon Swamp Forest, where a female was collected in 2011 from a deeply shaded, tiny pool by a small side stream. Teneral female collected in the same locality in 2010, earlier identified as *Amphicnemis*, represents the earliest record of the species in Singapore.

Etymology Generic epithet combines the Greek for 'to stretch out' (*teino*) and 'base' (*basis*), referring to rather long petiole of wing. Specific epithet, from the Latin *crypticus*, refers to the damselfly's cryptic colouration and behaviour.

Distribution Singapore, Peninsular Malaysia and Borneo.

National Conservation Status Critically Endangered; Restricted and Very Rare.

IUCN Red List Status Least Concern.

Larva Unknown.

Robin Ngiam

Female collected by Rory Dow from Nee Soon Swamp Forest in 2011.

Robin Ngiam

Teneral female collected by Cheong Loong Fah from Nee Soon Swamp Forest in 2010.

Red-tailed Sprite *Teinobasis ruficollis* (Selys, 1877)

Size HWL: 19–20mm; TBL: 36–38mm

Description Fairly small and slender damselfly with striking colours. Eyes brown and green. Male's thorax deep red with metallic silvery-blue stripe on dorsum of synthorax. Abdomen mostly dark except for segments 8–10 and appendages, which are deep red. Female similar to male but slightly paler. May be mistaken for young female Will-o-wisp (p. 90) in poor light, but latter has proportionately longer abdomen and lacks dark dorsal stripe on thorax.

Habitat & Habits Found in marshes and shallow pools in shaded swampy forests. Forest-dependent species that breeds at well-shaded, shallow pools with rich layer of leaf litter. Perches very low near the water's surface or in nearby vegetation, its shade-loving habit rendering it inconspicuous despite its colours.

Presence in Singapore Recorded in several locations, but almost always in very low densities: the Central Catchment and Bukit Timah Nature Reserves, Sungei Buloh Wetland Reserve, the Western Catchment, Admiralty Park, Thomson Nature Park, Windsor Nature Park, Pulau Semakau, Pulau Ubin and Pulau Tekong. Described from a specimen collected by Wallace in Singapore.

Etymology Specific epithet combines the Latin for 'red' (*rufus*) and 'necked' (*collis*), and probably refers to red prothorax.

Distribution Singapore, Peninsular Malaysia, Borneo and Sumatra.

National Conservation Status Near Threatened; Widespread but Rare.

IUCN Red List Status Near Threatened.

Larva Unknown.

Marcus Ng

A male. Despite their rich colours, these slender damselflies can be hard to spot in their dimly lit swampy habitats.

This shade-loving insect usually perches very low and seldom ventures far from the shadows.

Pair in wheel by a shaded forest pool in Admiralty Park.

Devadattidae (Grisettes)

This is a small and primitive (ancient in evolutionary lineage) family, with one genus and at least 13 species, found only in mainland and island Southeast Asia. Grisettes are moderately large, stocky and inconspicuous damselflies that are restricted to small streams in mature forests with dense canopies. Their wings are stalked, bear pterostigmata and are often darkened at the tips.

The generic epithet was coined in 1890 by Kirby. *Devadatta* means 'god-given' in Sanskrit, an ancient Indian language, and is also the name of a prominent Buddhist monk and cousin of Gautama Siddhartha, the founder of Buddhism. The family Devadattidae was named in 2013 when *Devadatta* was separated from the Amphipterygidae, which is now regarded as a purely neotropical group. *Gris* is French for 'grey' and 'grisette' refers to a drably clad young woman.

Marcus Ng

Male perched just above a small cascading portion of a hilly stream.

Malayan Grisette *Devadatta argyoides* (Selys, 1859)

Size Male HWL: 27–28mm; TBL: 40–42mm

Description Moderately large, dull-coloured damselfly with robust, boxy thorax. Eyes greyish. Thorax and abdomen mostly silvery blue-grey, with light brown banding, sometimes obscured, on abdominal segments 2–8. Wings stalked and hyaline, but with slightly darkened tips, and may reflect blue-purple sheen under certain light. Pterostigmata silvery-grey and teardrop shaped. Female similar to male but paler. May be confused with the Blue-spotted Flatwing (p. 55) in poor light, especially when wings are held open, but has a more strongly built (boxier) thorax and about eight antenodal crossveins in each wing. Flatwing has a more slender thorax and just two antenodal crossveins.

Habitat & Habits Found at well-shaded streams in dense, often hilly forests. Inconspicuous and usually perched very low. Males guard small cascading portions of forest streams, keeping very close to the water and avoiding sunlit spots. Both sexes may also be encountered at shaded parts of trails or around large fallen trunks not far from streams. Seldom flies far when disturbed, and often holds wings open for a short while right after landing.

Presence in Singapore Recorded in Bukit Timah Nature Reserve, Dairy Farm Nature Park, Rifle Range forest and Windsor Nature Park. Originally described as *Tetraneura argyoides* from a specimen collected in Singapore by Wallace.

Etymology See family description opposite for etymology of genus. Specific epithet probably derived from *argyros*, Greek for 'silver'; Selys's original 1859 description stated that the head was 'metallic steel in front and above' (translated from the original French).

Distribution Singapore, Peninsular Malaysia, southern Thailand and Sumatra.

National Conservation Status Endangered; Restricted and Uncommon.

IUCN Red List Status Least Concern.

Larva Fairly squat and hides in debris in shallow riffles. Caudal gills sturdy and pyramidical.

Marcus Ng

Female showing its thicker abdomen with a strong ovipositor.

Marcus Ng

Both sexes often hold their wings open for a few seconds after settling on a new perch.

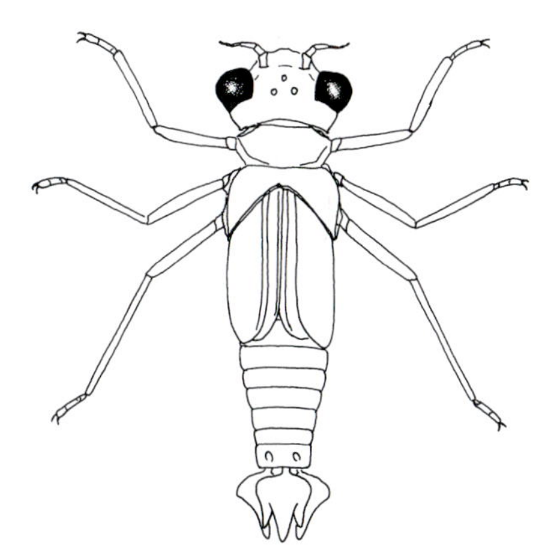

Albert G. Orr

Drawing of the larva, with its three pyramidical caudal gills.

Euphaeidae (Satinwings)
The satinwings are medium-sized to fairly large damselflies with moderately broad wings, which are usually unstalked and often brightly coloured or iridescent. The wings have pterostigmata and numerous antenodal crossveins. The legs are markedly shorter and the body is stockier than that of demoiselles. The family contains at least 77 species in nine genera in Asia and eastern Europe. Most species are restricted to streams and swamps with flowing water, usually in forested areas.

The etymology is unclear, but Sélys, who coined the genus *Euphaea* in 1840, may have combined *eu* (Greek for 'good' or 'true') with *phaûos*, Greek for 'brilliant', which aptly describes many members of the family.

Marcel Silvius

Male Euphaea variegata *Rambur, 1842 in Bali, showing the extensively coloured and iridescent wings typical of the genus.*

BLACK VELVETWING *Dysphaea dimidiata* Selys, 1853

Size HWL: 29–32mm; TBL: 44–48mm

Description Large, heavily built damselfly that may be mistaken for a true dragonfly in its habitat. Eyes black. Male has all-black body. Basal half (up to around nodus) and tips of both wings dark, showing purple-blue sheen under good light. Female has black and brownish-yellow markings, and hyaline wings with brownish tint.

Habitat & Habits Occurs at large, clear, fast-flowing forest streams with rocky or pebbly bottoms. Where present, male often occupies a sunny spot on a stone, log or branch, holding the wings wide open and even depressed, resembling a true dragonfly from afar. From this perch, male makes short, flitting patrol flights before returning to the original position, but may chase rivals for up to 100m. Females seldom seen by water, preferring to forage in the forest or canopy except when breeding.

Presence in Singapore Last collected locally by Wallace in 1854, whose specimen was described as *Dysphaea dimidiata limbata* by Sélys in 1859. Its preferred habitat has since vanished from the island with the loss or canalization of larger forest streams.

Etymology In Latin, *dys-* means 'abnormal' or 'impaired', while *dimidiata* means 'diminished'. The binomial may thus refer to the partially coloured wings. Sélys may have coined the genus *Dysphaea* to distinguish this species from other known members of this group then, such as *Euphaea variegata*, Rambur, 1842, which have almost fully coloured wings.

Distribution Singapore (formerly), Sundaland and southern Thailand.

National Conservation Status Extinct.

IUCN Red List Status Least Concern.

Larva Unknown but probably similar to larvae of *Euphaea* species (p. 129).

Tang Hung Bun

Male showing the iridescent upper surface of the wings. Photo taken in Johor, Malaysia.

Erland Refling Nielsen

Male with wings held closed. Photo taken in Pahang, Malaysia.

Matti Hämäläinen

Female showing its hyaline wings and striped markings. Photo taken in Johor, Malaysia.

Erland Refling Nielsen

Pair in tandem near the edge of a large forest stream. Photo taken in Pahang, Malaysia.

Blue-sided Satinwing *Euphaea impar* Selys, 1859

Size HWL: 23–25mm; TBL: 34–38mm

Description Medium-sized damselfly with distinctive colours. Eyes black and blue-grey, paler in female than in male. Male's thorax has bright blue sides and black dorsum. Abdomen and appendages black. Wings unstalked. Hindwing slightly shorter than forewing and bears dark, non-iridescent patch on its distal portion. Males with hyaline wings have been seen a few times in Singapore, and may represent a rare form or younger individuals. Female has hyaline wings and olive-grey thorax.

Habitat & Habits Found at shaded, slow-flowing forest streams with rich fringing vegetation and sandy or muddy bottoms; also small streams in swamp forests. Forest-dependent damselfly seldom found far from shaded streams. Usually perched quite close to the water, but on brighter days males may occupy higher and sunlit perches over the water, at times with wings half-open. Flight fluttery and usually brief, often ending at original perch. Dark wing-patch is thought to play a role in intraspecific contests. Females less often seen than males, but may forage in clearings near trails close to streams. Courtship rudimentary, with male seizing female after a brief chase. Oviposition takes place in submerged vegetation, with female crawling down a stem or plunging into the water at high speed, and remaining underwater for as long as 30 minutes.

Presence in Singapore Recorded in the Bukit Timah and Central Catchment Nature Reserves and adjacent nature parks such as Windsor Nature Park.

Etymology Refer to the family description for the possible etymology of *Euphaea*. In Latin, *impar* means 'odd' or 'uneven'. This specific epithet may have been given as most other known *Euphaea* species have dark-coloured bodies and extensively coloured, iridescent wings. A very similar species, *E. ameeka* Van Tol & Norma-Rashid, 1995, with largely blue thorax but lacking dark wing-tips, occurs in Brunei and parts of Sarawak.

Distribution Singapore, Peninsular Malaysia, Borneo, Sumatra and southern Thailand.

National Conservation Status Least Concern; Widespread and Common.

IUCN Red List Status Least Concern.

Larva Typical of family. Robust with seven rows of long gills on sides of abdomen and three leaf-like caudal gills. Occurs under leaf packs in areas with a swift current.

Marcus Ng

The bright blue thorax is the most conspicuous part of the male.

Marcus Ng

The olive-grey female may forage along shady trails near streams.

Marcus Ng

The male may hold its wings slightly open while basking.

Marcus Ng

Female basking with wings held open.

Marcus Ng

Young male with paler colours and hyaline wings.

Robin Ngiam

The rather stout larva, showing the leaf-like caudal gills.

Lestidae (Spreadwings)

The spreadwings are medium-sized to large but lightly built damselflies that perch with their abdomens pointing downwards at an angle and wings partially or fully spread open. The legs are moderately long. The wings are stalked and usually hyaline, with long pterostigmata. There are more than 150 species in nine genera worldwide.

Spreadwings are often associated with marshy habitats, hence their other moniker, 'reedlings', or with forested swamps. *Lestes* is Greek for 'robber', alluding perhaps to the insects' voracious appetite, especially that of the larvae, which are more active swimmers than the larvae of pond damsels and featherlegs. The name may also be derived from *leste* (French for 'nimble').

Marcus Ng

Orolestes selysi McLachlan, 1895, a large spreadwing found in Thailand, Indochina and southern China.

CRENULATED SPREADWING *Lestes praemorsus*
Hagen in Selys, 1862

Size Male HWL: 20–21mm; TBL: 38–42mm

Description Medium-sized, slender damselfly with azure-blue eyes and widely held wings. Male's thorax pruinosed grey-blue with small dark spots. Dorsum of synthorax has dark, sometimes faint band with crenulated edges. Abdomen darkish with pale banding; segments 9–10 may have grey-blue pruinescence. Upper appendages mostly white, lowers black. Abdomen-tip slightly upturned. Female similar but duller in colour – more greyish than blue. Local populations currently thought to be subspecies *decipiens*, but the entire *praemorsus* group is due for a proper taxonomic review with molecular analysis.

Habitat & Habits Occurs around medium-sized to large ponds, lakes, marshes and reservoirs, especially those with shallow banks lined by rich marginal vegetation (for example *Eleocharis* species and other sedges). Adults cling to low vegetation with wings held open and abdomen pointing downwards in typical lestid pose. After mating, male remains in tandem while female inserts eggs into plant stems.

Presence in Singapore Fairly common around MacRitchie Reservoir (a stronghold). Also recorded in Upper Seletar Reservoir Park, Singapore Quarry in Dairy Farm Nature Park (most probably extirpated there), Bukit Timah Saddle Club, the Western Catchment, Pulau Tekong and Pulau Ubin.

Etymology See family description for etymology of *Lestes*. In Latin, *praemorsus* means 'to bite off', perhaps referring to the insect's predatory behaviour.

Distribution Tropical Asia and New Guinea.

National Conservation Status Least Concern; Widespread but Uncommon.

IUCN Red List Status Least Concern.

Larva Typical of damselflies, with elongated body and three flattened caudal gills. Central gill about one and a half times longer than other two. Found among aquatic vegetation.

Marcus Ng

Male showing the characteristic perching and wing posture of the genus.

The tip of the abdomen is turned slightly upwards when perched.

Crenulated Spreadwings in tandem in a reedbed.

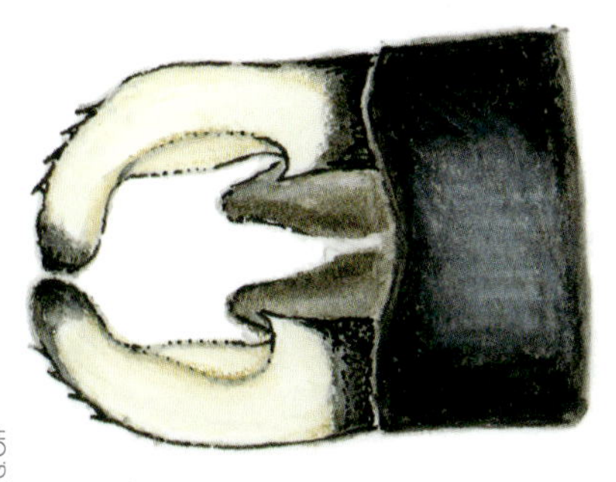

Lateral (top) and dorsal views of the male anal appendages.

The male remains in tandem with its mate while the female places its eggs into submerged plant matter.

Great Spreadwing *Orolestes wallacei* (Kirby, 1889)

Size HWL: 29–32mm; TBL: 55–60mm

Description Large but slender damselfly with green eyes. Male's thorax light green with darker bands; abdomen mostly dark with segments 8–9 light blue. Appendages dark. Wings hyaline, with much longer pterostigmata than those of other local lestids. Female duller and without blue on abdomen.

Habitat & Habits Occurs in swampy forests. Perches with wings outspread and abdomen hanging downwards. Where present, found around shaded, marshy areas and small pools in dense, swampy forests, perched low amid tangled vegetation. Flies up to higher perch when disturbed.

Presence in Singapore The last Singapore record was supposedly a female collected by Ridley in the late nineteenth century, said to be in the British Museum (now the Natural History Museum, London) collection. This was mentioned by Laidlaw (1902), who did not actually examine the specimen.

Etymology *Oros* means 'mountain' in Greek. Genus coined in 1895 by Robert McLachlan for *O. selysi*, a species discovered in the Himalayas (though not all *Orolestes* species are restricted to the highlands). Specific epithet honours Wallace, who collected the type specimen in Sarawak.

Distribution Sundaland.

National Conservation Status Extinct.

IUCN Red List Status Least Concern.

Larva Fairly typical of family – found in shallow forest ponds.

Erland Refling Nielsen

Dorsal view of a male photographed in Taman Negara, Malaysia.

Erland Refling Nielsen

Lateral view of the male.

Erland Refling Nielsen

Female photographed in Taman Negara, Malaysia. Great Spreadwings often perch very low amid dense vegetation, making them difficult to spot (and even harder to photograph) despite their size.

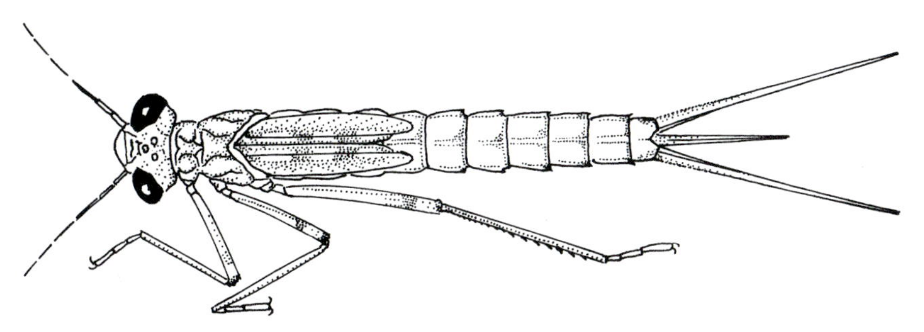

Albert G. Orr

Drawing of the larva.

SLENDER SPREADWING *Platylestes heterostylus* Lieftinck, 1932

Size Male HWL: 20–21mm; TBL: 41–42mm

Description Medium-sized, lightly built spreadwing with green eyes and muted greenish colours. Thorax light green with scattered dark spots. Synthorax lacks dark dorsal band, distinguishing this species from the Crenulated Spreadwing (p. 132). Wings hyaline with short pterostigmata. Abdomen long and slender, with pale, robust appendages.

Habitat & Habits Found in open marshy areas and swampy forests.

Presence in Singapore Recorded in Bukit Timah in the 1960s–'70s, when three specimens were sent by Murphy to Vancouver Museum, Canada for verification. Last recorded locally by Dennis Paulson, who found one male at the west end of MacRitchie Reservoir in 1980.

Distribution Sundaland.

National Conservation Status Critically Endangered; Restricted and Very Rare.

IUCN Red List Status Data Deficient.

Larva Unknown.

Anthony Quek

Male photographed in Endau-Rompin National Park, Malaysia.

Mandy Lee

Male photographed in Langkawi, Malaysia.

Platycnemididae (Featherlegs)

These small to fairly large damselflies of very slender to medium build have fairly long legs and narrow wings. Some species have expanded leg segments that are used in territorial or courtship displays; this feature gave rise to the family name: *platy* is Greek for 'broad' and *cnemis* means 'shin'. The wings are stalked, hyaline in local species, and bear short pterostigmata. Most local featherlegs are restricted to slow-flowing streams, ponds and swampy areas in shaded forests and forest edges.

The family includes more than 460 species across the Old World. The Shorttail (formerly Coenagrionidae: Argiinae) and Old World threadtails (formerly Protoneuridae, Disparoneurinae) are now reassigned to Platycnemididae.

Marcus Ng

Pseudocopera ciliata (*Selys, 1863*), *a featherlegs found in mainland tropical Asia with visibly expanded leg segments.*

Marcus Ng

Male Yellow Featherlegs photographed in Thailand, showing its colourful and moderately expanded legs.

White-tailed Sylvan *Coeliccia albicauda*
(Förster in Laidlaw, 1907)

Size HWL: 22–24mm; TBL: 41–44mm

Description Medium-sized, long-legged damselfly with fairly slender build. Male's eyes black and blue. Synthorax black with four irregular blue spots on dorsum and blue stripes on lower sides. Abdomen dark. Segment 10 and appendages very pale. Female's eyes dark brown and light green. Synthorax blackish with blue-green dorsal stripes and broader yellowish stripes on sides. Dorsum of prothorax pale – compare with dark dorsum of the female Twin-spotted Sylvan (p. 140). Abdomen dark brown with paler markings on segments 8–9.

Habitat & Habits Found around small streams and seepages in shaded forests. Perches on low vegetation in shaded corners of small streams and seeps with abdomen pointing slightly downwards and wings held half-open, but never as wide as in spreadwings.

Presence in Singapore Locally recorded only once, in Nee Soon Swamp Forest in 1994.

Etymology Generic epithet may be derived from *koilos*, Greek for 'hollow' or 'concave', referring to female prothorax of some members of genus, which is indented at the rear margin. Specific epithet combines *albus*, Latin for 'white', and *cauda* ('tail').

Distribution Singapore, Peninsular Malaysia, Thailand and Myanmar (probable).

National Conservation Status Critically Endangered; Restricted and Very Rare.

IUCN Red List Status Least Concern.

Larva Unknown but should be typical of the genus (see the Telephone Sylvan, p. 142). Caudal gills may be petiolated. Probably found in leaf packs in streams.

Marcus Ng

The white appendages easily distinguish the male White-tailed Sylvan from other local members of the genus.

Marcus Ng

Dorsal view of the male, showing the spots on the dorsum.

Oleg Kostein

Female photographed in Thailand.

Twin-spotted Sylvan *Coeliccia didyma* (Selys, 1863)

Size HWL: 22–24mm; TBL: 41–46mm

Description Medium-sized, fairly lightly built damselfly with long legs. Very similar to the White-tailed Sylvan (p. 138), but spots on dorsum of synthorax are more regular and elongated in shape. Also, abdominal segments 9–10 are largely blue, while appendages are black, contrasting with pale appendages of White-tailed. Female has dark brown and light green eyes. Synthorax dark with pale yellow dorsal stripes, straighter than in White-tailed, and broader yellow stripes on sides.

Habitat & Habits Found at seepages in forests and forest edges. Elsewhere in region, where it is fairly common in suitable habitat, can be found on low vegetation in forests and forest edges, close to small streams and seeps.

Presence in Singapore Locally recorded only once, in Chestnut Forest in 1993.

Etymology Specific epithet is Greek for 'twin' and was an appellation for the twin deities Apollo and Artemis. It may refer to the closely placed pair of larger spots on dorsum of synthorax.

Distribution Mainland Southeast Asia to India.

National Conservation Status Critically Endangered; Restricted and Very Rare.

IUCN Red List Status Least Concern.

Larva Unknown.

Marcus Ng

Male photographed at Fraser's Hill, Malaysia, showing the blue terminal segments and dark appendages.

Female photographed in Kaeng Krachan National Park, Thailand.

Pair in tandem by a tiny seep at a forest edge near Fraser's Hill, Malaysia.

TELEPHONE SYLVAN *Coeliccia octogesima* (Selys, 1863)

Size HWL: 21–22mm; TBL: 40–42mm

Description Medium-sized, slender damselfly with long legs and unique markings. Male's eyes dark brown and light blue. Synthorax dorsum has paired light blue markings resembling the handset of an old-fashioned telephone. Abdomen blackish, except for segments 9–10, which are largely blue. Appendages black. Female has brown and light green eyes. Synthorax with similar 'telephone-handpiece' markings, but these may be less well defined or even fragmented into paired spots. Abdomen brownish, darker at tip.

Habitat & Habits Found at small streams, pools and seepages in shaded forests and swampy forests. Perches on low vegetation with wings half-open. May be locally abundant in swampy areas, but both sexes, especially females, can also be encountered foraging along forest trails some distance from water.

Presence in Singapore Mostly confined to the Central Catchment Nature Reserve and adjacent nature parks such as Windsor Nature Park and Thomson Nature Park. Also at Chestnut Nature Park and Rifle Range forest. Described from a specimen collected by Wallace in Singapore in 1854.

Etymology Specific epithet means 'eightieth' in Latin. Etymology unclear, but Sélys wrote of the female: 'The blue markings on the dorsum are less widely separated from one another than in the male, forming on either side almost a figure 8.'

Distribution Singapore, Peninsular Malaysia and Sumatra.

National Conservation Status Vulnerable; Restricted but Common.

IUCN Red List Status Least Concern.

Larva Similar to those of coenagrionoid larvae: small, lightly built and long legged. Occurs in leaf litter in forest pools and streams.

Marcus Ng

Lateral view of the male, showing the distinctive telephone handpiece-like dorsal markings.

Marcus Ng

Female with fairly well-defined dorsal markings.

Marcus Ng

Dorsal view of the male, perched very low in a shaded swampy forest.

Marcus Ng

The female's dorsal markings are less well defined or even fragmented in some individuals.

Marcus Ng

Pair in tandem by a small shaded pool in swampy forest.

Yellow Featherlegs *Copera marginipes* (Rambur, 1842)

Size HWL: 16–18mm; TBL: 35–38mm

Description Smallish, slender damselfly with fairly long and brightly coloured legs. Eyes dark brown with pale bands. Male's thorax black, with greenish-yellow streaks on sides. Legs bright yellow, with expanded femora and tibia. Abdomen dark, with segments 8 (partially), 9–10, and appendages white. Female similar but with paler markings and lacking white on abdomen-tip. Younger adults almost entirely white. This 'ghost' form – a feature of the genus – darkens with age.

Habitat & Habits Found around clear streams, sluggish channels and shallow swamps in forests or near forest edges. Perches low but is fairly conspicuous due to bright colours of mature adults as well as younger 'ghost' forms. Wings may be held slightly open when perched. When disturbed, may hover briefly before settling on the same perch or very close by.

Presence in Singapore Very common elsewhere in the region, but less so in Singapore. Recorded in several locations with suitable habitats, such as the Central Catchment Nature Reserve, Windsor and Thomson Nature Parks, the Western Catchment, Bukit Batok, Mandai, Sembawang Hot Spring, Tengah and Clementi Forest.

Etymology Unclear, although in South American Spanish, *copera* refers to a waitress. Kirby coined the genus in his 1890 *A Synonymic Catalogue of Neuroptera Odonata, or Dragonflies* without explanation. Originally described as *Platycnemis marginipes*.

Distribution Widespread in tropical Asia.

National Conservation Status Least Concern; Widespread and Common.

IUCN Red List Status Least Concern.

Larva Fairly typical of coenagrionoid larvae, but caudal gills bear long filaments at margins (see the Variable Featherlegs, p. 146). Occurs in root masses and debris in slow-flowing streams.

Marcus Ng

The mature male is conspicuous even in its shaded habitats, due to its bright yellow legs.

Marcus Ng

Mature female photographed in Thailand.

Marcus Ng

A 'ghost' form young female photographed in Gopeng, Malaysia.

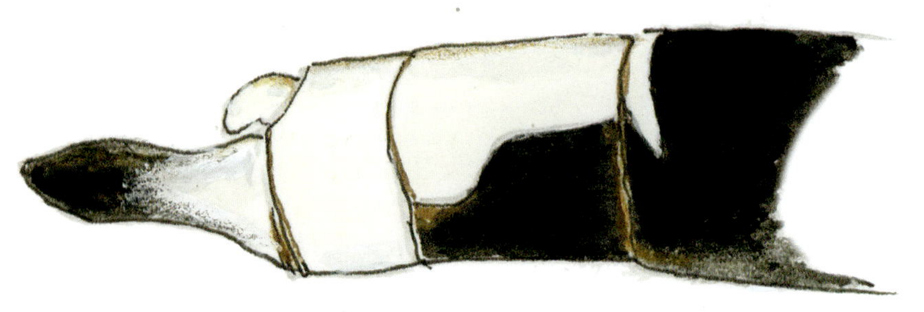

Albert G. Orr

Drawing of the male's abdomen-tip and anal appendages.

Variable Featherlegs *Copera vittata* (Selys, 1863)

Size HWL: 16–17mm; TBL: 34–36mm

Description Small dark damselfly, more lightly built than the Yellow Featherlegs (p. 144). Eyes greenish with contrasting bands. Male's thorax black with greenish-yellow variegation. Legs yellow with expanded femora and tibia; elsewhere in region, males may have orange-red or black legs. Abdomen mostly black except for segment 10, which is white. Upper appendages white, lowers longer and black. Female has darker markings and less brightly coloured legs. Like in Yellow, young adults have a pale 'ghost' form. Several forms occur in region and may represent a species complex in need of taxonomic revision.

Habitat & Habits Found at sluggish channels and shallow pools in swampy forests. Younger 'ghost' forms may forage in undergrowth further from the water. Perches very low and keeps to the shade. Less conspicuous than Yellow due to its daintier build and darker colours.

Presence in Singapore Rare and local in Singapore. Recorded in the Western Catchment, Admiralty Park, Nee Soon Swamp Forest, Pulau Ubin and Pulau Tekong.

Etymology Specific epithet comes from *vittatus*, which means 'banded' in Latin.

Distribution Tropical Asia.

National Conservation Status Vulnerable; Restricted and Rare.

IUCN Red List Status Least Concern.

Larva Similar to that of the Yellow Featherlegs, but caudal gills more slender at bases and paler with dark bands.

Marcus Ng

The thoracic markings and abdomen-tip (less white) distinguish the Variable Featherlegs from the slightly larger Yellow Featherlegs.

Marcus Ng

Pale 'ghost' form of the young male. Photo taken in Gopeng, Malaysia.

Keith Wilson

Mature female. Photo taken in Endau-Rompin, Malaysia.

Robin Ngiam

Variable Featherlegs in tandem.

Robin Ngiam

Dorsal view of the larva, showing the filamentous caudal gills.

SHORTTAIL *Onychargia atrocyana* Selys, 1865

Size HWL: 17–18mm; TBL: 30–32mm

Description Smallish damselfly with somewhat stout build and distinctively short abdomen. Legs fairly long, lined with dense spines. Male has black and dark blue eyes; thorax completely dark purplish-blue. Abdomen and appendages black. Female has black and light green eyes. Synthorax dark with light yellowish-white stripes. Younger males similar to female, turning darker as they mature.

Habitat & Habits Found around leafy pools and marshes in forests and forest edges. Tends to perch fairly high, up to several metres from the ground, on shrubs and low trees, coming down to the water mainly to breed. Towards midday, pairs in tandem or wheel may be encountered on low vegetation close to water, but they retreat to higher foliage when disturbed.

Presence in Singapore Recorded in many locations, including the Central Catchment Nature Reserve, adjacent nature parks such as Windsor Nature Park and Thomson Nature Park, Admiralty Park, Kent Ridge Park, Kranji Marshes, Bukit Brown, Pulau Ubin and the Singapore Botanic Gardens (Plant House and Symphony Lake). Singapore, where Wallace collected the first specimens, is the type locality.

Etymology Generic epithet probably combines *onycho-*, Greek for 'claw', with *Argia*, a genus of American pond damselflies formerly thought to be closely related to this species. Specific epithet combines the Latin for 'black' (*atrum*) and 'blue' (*cyanos*). Formerly thought to be a pond damselfly (Coenagrionidae, Argiinae), but following genetic studies (Dijkstra et al., 2013) has been reassigned to new subfamily, Onychargiinae, under the Platycnemididae.

Distribution Widespread in tropical Asia.

National Conservation Status Least Concern; Widespread but Uncommon.

IUCN Red List Status Least Concern.

Larva Typical coenagrionoid in appearance. Caudal gills leaf shaped and petiolated with mottled markings; median gills longer than lateral gills.

Marcus Ng

In the field, the mature male comes across as a smallish, very dark damselfly with a shortish abdomen.

Marcus Ng

The younger male resembles the female, developing darker hues as it matures.

Marcus Ng

Female, showing the shortish abdomen of the species.

Marcus Ng

Mating pairs will come down low to perch above small pools in marshy forest.

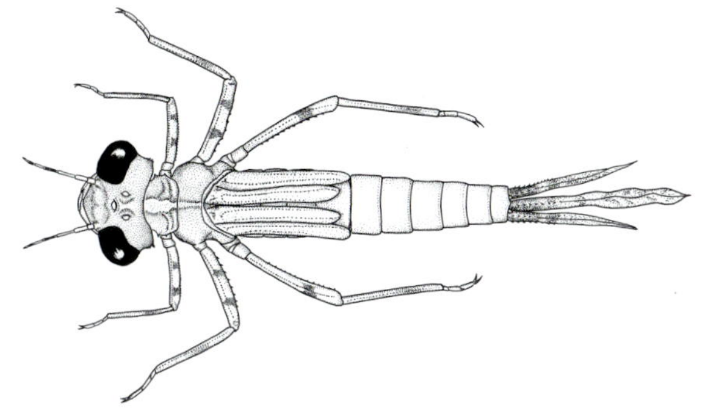
Albert G. Orr

Drawing of the larva.

Collared Threadtail *Prodasineura collaris* (Selys, 1860)

Size HWL: 17–18mm; TBL: 33–35mm

Description Small, very slender damselfly with black and greyish-blue eyes. Male has blue band on head between eyes. Rear lobe of prothorax blue, forming eponymous 'collar'. Dorsum of male's synthorax has blue foliar (leaf-shaped) marking that tapers to a point. Abdomen blackish, appendages whitish-blue. Female's synthorax has very short and incomplete pale blue stripe on dorsum. Abdomen blackish and lacking blue at tip.

Habitat & Habits Found at sluggish forest streams in shaded forests and swampy forests. Perches low, close to water, sometimes with wings slightly open. Sexual activity peaks towards midday. Pairs in wheel perch on low vegetation near slow channels and pools, before descending to the water. After mating, male remains in tandem while female oviposits in shallow root masses. Both sexes may bask at forest edges, especially in the late afternoon.

Presence in Singapore Restricted to the Central Catchment Nature Reserve, Windsor and Thomson Nature Parks. Also found in Bukit Brown.

Etymology Generic epithet is an anagram of *Disparoneura* ('separately veined'), a closely related genus with similar wing venation. Specific epithet, meaning 'collared', may refer to the fully blue posterior lobe of the male's prothorax. Alternatively, it may refer to the rear lobe of the female's prothorax, which differed in shape from the lobes of other known species when Sélys described it in 1860.

Distribution Sundaland, Thailand and southern Myanmar.

National Conservation Status Vulnerable; Restricted and Uncommon.

IUCN Red List Status Least Concern.

Larva Fairly broad, angulated head, shortish body and proportionately long, dark banded legs. Tiny spines along margin of caudal gills, the bases forming a black spot. Found in root masses in streams.

Marcus Ng

Male showing the blue 'collar' on the prothorax and the pale appendages.

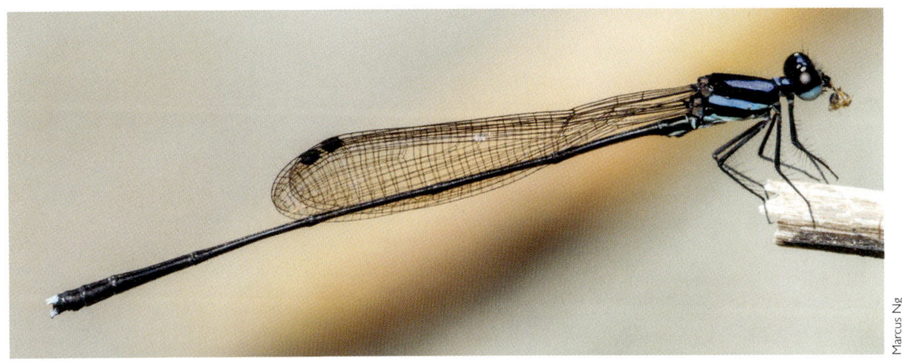

Marcus Ng

Lateral view of a male. The dorsal stripe appears purplish in this photograph due to the use of a flashgun.

Marcus Ng

Female showing the very short dorsal stripe and dark abdomen.

Marcus Ng

Pair in wheel.

Albert G Orr

Drawing of the male anal appendages.

ORANGE-STRIPED THREADTAIL *Prodasineura humeralis* (Selys, 1860)

Size HWL: 17–18mm; TBL: 34–37mm

Description Smallish, dark and lightly built damselfly with very slender abdomen. Eyes reddish-brown, with contrasting banding in female. Male's thorax dark brown, with thin and incomplete brownish-orange dorsal stripe and broader, pale orange stripe on sides. Abdomen and appendages black. Thoracic stripes and upper appendages whitish in young males. Female's thorax blackish with pale dorsal stripes, which are longer and slightly thicker than male's. Abdomen darkish with slight banding. Banding on eyes distinguishes female from female Cresent Threadtail (p. 154). Some taxonomists consider this to be a subspecies of *P. verticalis*.

Habitat & Habits Found around forest streams with moderate- to fast-flowing water; also reservoir inlets with dense overhanging vegetation. Young males have been observed in abundance along reservoir edges and forest trails. Mature males perch very close to the water, and when disturbed, hover briefly before alighting on the same perch or close by. Females may be found amid streamside vegetation or along nearby trails.

Presence in Singapore Formerly known only from Mandai and Upper Peirce, but has since spread to many other sites. Recorded in the Central Catchment Nature Reserve, Windsor Nature Park, Dairy Farm Nature Park, Chestnut Forest, Bukit Batok Nature Park and Bukit Brown, among others.

Distribution Singapore and Peninsular Malaysia.

National Conservation Status Least Concern; Widespread and Common.

IUCN Red List Status Least Concern.

Larva Similar in general appearance to larva of the Collared Threadtail (p. 150).

Marcus Ng

Male showing its orange thoracic markings. Mature males usually perch very low at fast-flowing streams.

Marcus Ng

Young male, with pale thoracic markings, appendages and banding on the eyes.

Marcus Ng

Young female with paler colours.

Marcus Ng

A mature female. The banding on the eyes and darker abdomen distinguish it from the female Crescent Threadtail.

INTERRUPTED THREADTAIL *Prodasineura interrupta* (Selys, 1860)

Size HWL: 15–18mm; TBL: 31–35mm

Description Small, lightly built damselfly. Distinguished from other blue threadtails by thin, straight-sided and incomplete (interrupted) blue stripe on dorsum of male's synthorax. Abdomen and appendages black, with blue markings on dorsum of segments 9–10 and upper appendages. Female has pale blue bands on thorax.

Robin Ngiam

Dorsal view of a male, showing the incomplete, straight-sided dorsal stripes and blue on abdominal segments 9–10.

Tang Hung Bun

Lateral view of a male.

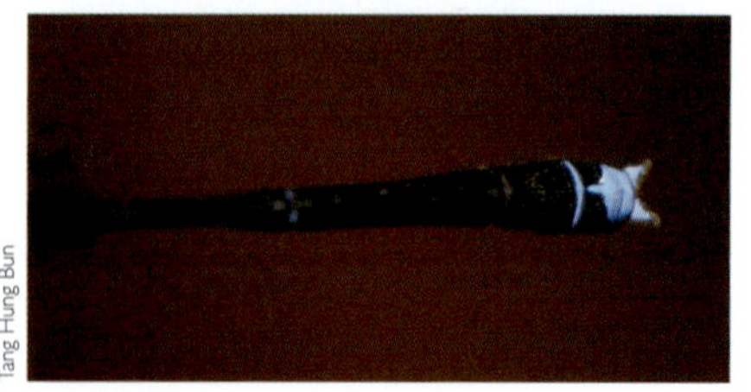

Tang Hung Bun

The male Interrupted Threadtail has more extensive blue on the terminal segments than the Collared Threadtail.

Habitat & Habits Found around sluggish streams and rivulets in shaded swampy forests.

Presence in Singapore Described from a specimen collected by Wallace in Singapore, supposedly from Bukit Timah. Presently confined to Nee Soon Swamp Forest. Rarest of Singapore's threadtails.

Etymology Specific epithet probably refers to truncated dorsal stripe on synthorax.

Distribution Singapore, Peninsular Malaysia and Sumatra.

National Conservation Status Critically Endangered; Restricted and Uncommon.

IUCN Red List Status Least Concern.

Larva Unknown, but should be typical of genus.

Tang Hung Bun

Head-on view of a male, showing the incomplete thoracic stripes and lack of a blue 'collar' (unlike in the Collared Threadtail).

Albert G. Orr

Drawing of the male's abdomen-tip and anal appendages.

CRESCENT THREADTAIL *Prodasineura notostigma* (Selys, 1860)

Size HWL: 17–19; TBL: 34–37mm

Description Small, lightly built damselfly with dark and slender abdomen. Eyes black and greyish-blue. Dorsum of male's synthorax has leaf-like blue mark that tapers at the end. Differs from the Collared Threadtail (p. 150) by lack of blue band between eyes and blue 'collar'. Also, appendages are black, not blue. Female has reddish-brown abdomen and very thin but fairly complete light blue dorsal stripe on synthorax.

Habitat & Habits Found around slow-flowing forest streams. May occur with Collared, but found in slightly more exposed areas and prefers streams to swamps. Males perch at streams very close to the water. Both sexes may bask and feed along forest trails near streams. Sexual behaviour and oviposition similar to that of Collared.

Presence in Singapore Recorded in various locations, including the Central Catchment and Bukit Timah Nature Reserves, Windsor Nature Park, Rifle Range forest and Bukit Batok Nature Park. Described from a specimen collected by Wallace in Singapore.

Etymology Specific epithet appears to combine the Greek for 'back' (*noto-*) and 'mark' (*stigma*). Sélys, who described the species as *Alloneura notostigma*, noted *une tache étroite oblongue au 2° segment* ('a narrow oblong mark on the second segment') of the abdomen.

Distribution Singapore, Peninsular Malaysia, Sumatra and Borneo.

National Conservation Status Least Concern; Widespread and Common.

IUCN Red List Status Least Concern.

Larva Typical of genus.

Marcus Ng

Mature male. The dark appendages readily distinguish this species from the Collared Threadtail.

Marcus Ng

The female's abdomen is often reddish-brown. The eye colours also distinguish it from the female Orange-striped Threadtail (p. 152).

Albert G. Orr

Drawing of the male's abdomen-tip and anal appendages.

Marcus Ng

The female has a thin but fairly complete dorsal stripe on the thorax.

Marcus Ng

Male hovering over a small forest stream. Note the lack of a blue band between the eyes and lack of a blue 'collar'.

Marcus Ng

Crescent Threadtails in wheel. Reproductive activity begins from mid-morning and continues until the early afternoon.

Platystictidae (Shadowdamsels)

These small to medium-sized, lightly built damselflies have very long and slender abdomens, especially the males. The wings are narrow, stalked and slightly falcate (hooked) at the tips. The pterostigmata are squarish. To date, more than 270 species in 10 genera have been described in Asia and the Neotropics, but the family is most diverse in tropical Asia. Shadowdamsels are weak fliers with poor dispersal powers, seldom wandering far from their breeding grounds. The genus *Drepanosticta*, with more than 125 species, contains many damselflies endemic to small islands or with very restricted distributions.

Shadowdamsels are secretive insects that typically lurk in the dimmest corners of forest streams, seepages and swamps. They are noticeable mainly by their blue abdominal tail-lights. Fraser (1933) observed of the family: 'Owing to their dull colouring, small size, and dark surroundings, they are remarkably inconspicuous during flight, and would be invisible were it not for the chain of white and blue spots on the abdomen seen to be moving stealthily about the dark recesses. The vivid blue identification marks on the terminal segments of the abdomen show up with remarkable conspicuousness even in the darkest retreats when the insect is at rest.'

The family name combines *platus,* Greek for 'broad', with *sticta*, which means 'marked' or 'spotted'. It was derived from *Platysticta,* a South Asian genus with prominent dark patches on the wing-tips.

Singapore Shadowdamsel *Drepanosticta quadrata* (Selys, 1860)

Size HWL: 20–22mm; TBL: 38–46mm

Description Medium-sized damselfly with very slender abdomen. Eyes dark brown. Male's synthorax blackish with pale stripe on lower side. Abdomen very long and thin, with blue on dorsum of segments 8–9. Segment 10 and appendages black. Wings narrow and hyaline, with slightly hooked tips. Female similar but with barely any blue on abdomen.

Habitat & Habits Found at deeply shaded streams in hilly forests; also swamp forests. As the name implies, prefers very dim spots by small streams and seepages, where it is nigh invisible except for the blue 'tail-light' on the abdomen. Perches very low, and when disturbed, hovers briefly before settling on vegetation nearby.

Presence in Singapore Recorded in the Bukit Timah and Central Catchment Nature Reserves, Dairy Farm and Windsor Nature Parks.

Etymology Generic epithet combines *drepano* (Latin for 'sickle-like', probably referring to the hooked wing-tips) and *sticta*, a common suffix for members of the family. Laidlaw, who coined the genus in 1917, noted the falcate wings. Specific epithet may refer to the squarish pterostigmata (*quadrata* means 'squarish' in Greek). Originally described as *Platysticta quadrata* based on specimens collected by Wallace in Singapore.

Distribution Singapore and Peninsular Malaysia (with certainty only from Johor).

National Conservation Status Vulnerable; Restricted but Common.

IUCN Red List Status Near Threatened.

Larva Unknown, but typical larvae of the genus are less elongated than most damselfly larvae, with a large, wedge-shaped head and tapering body. Caudal gills with long filaments. Found among plant debris or under stones in well-shaded small streams and seepages.

Marcus Ng

Against the blacks and browns of its preferred habitat, the blue 'tail-light' is the clearest sign of the male's presence.

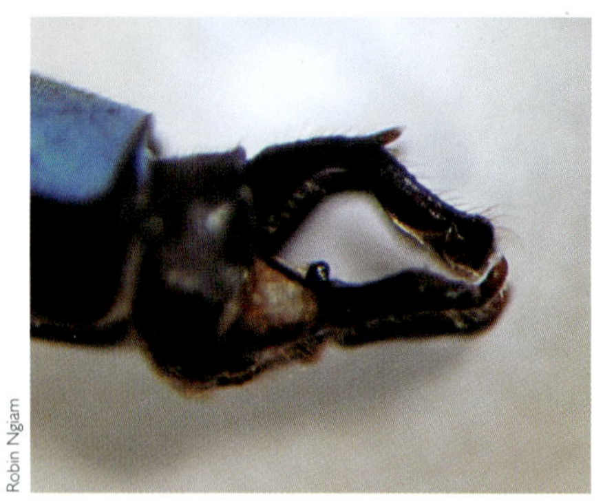
Robin Ngiam

Close-up of the male's anal appendages.

Tang Hung Bun

Female showing its thicker abdomen, which also lacks a blue 'tail-light'.

Cheong Loong Fah

Female ovipositing on a twig above a fast-flowing, well-shaded forest stream.

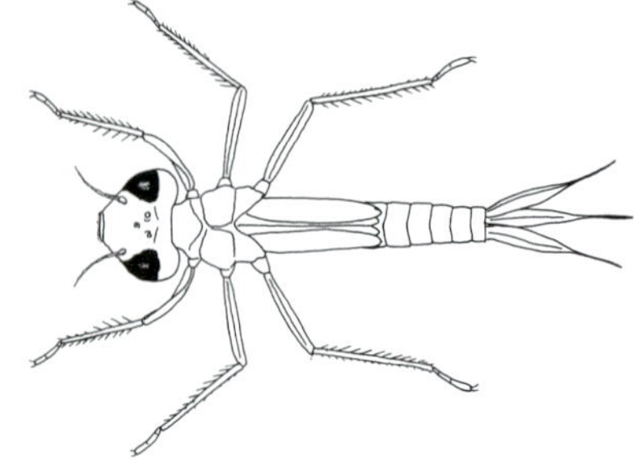
Albert G. Orr

Drawing of a typical Drepanosticta larva, showing its largish head.

ANISOPTERA

Aeshnidae (Hawkers)

Hawkers are large to very large, robust dragonflies with very long, slender abdomens that are often swollen at the base and constricted at the third segment, creating an hourglass-like shape. They are frequently greenish in colour, with very large green or blue eyes that meet broadly at the top of the head. The males' hindwings have an acute anal angle, except in the genus *Anax*. Female hawkers are the only true dragonflies in the region to have a functional ovipositor, which is used to insert eggs into soil or plant matter.

Worldwide, there are about 490 species of hawker (known as darners in North America). Some patrol ponds and other open waters, where they command attention with their bright colours and strong flight. Others are crepuscular, becoming active towards evening or before sunrise and hiding in dense vegetation by day. When perched, aeshnids invariably hang from a twig or leaf with their abdomens pointing downwards.

The etymology of the family name is confusing and complex. *Aeshna* is a large genus of hawkers found in the northern hemisphere. However, before Linnaeus, *Aeschna* (meaning unknown) referred to mayflies (Ephemeroptera) in England. In 1775, Fabricius used the generic name *Aeshna* to denote the larger true dragonflies, as opposed to the smaller skimmers or *Libellula*; the 'c' was omitted for unknown reasons – possibly due to a printing error or a slip of the pen.

Despite its spelling being ungrammatical in Greek, *Aeshna* is deemed a valid name by the International Commission of Zoological Nomenclature, a group that sets the rules and is the final arbiter for scientific names of animals, and the derived family name follows this spelling. However, other genera in the family, such as *Heliaeschna* and *Oligoaeschna*, spell this element with a 'c', following correct Greek convention.

Robin Ngiam

Male Aeshna mixta Latreille, 1805, or Migrant Hawker, from the UK. This colourful migratory species ranges from Europe and North Africa to China and Japan.

EMPEROR *Anax guttatus*
(Burmeister, 1839)

Size HWL: 52–56mm; TBL: 80–86mm

Description Very large, heavily built dragonfly with green and blue markings. Eyes, thorax and abdominal segment 1 green. Abdominal segment 2 and base of segment 3 blue. Segment 3 slightly constricted. Remaining segments dark with pale spots along sides. Wings hyaline but hindwings of older individuals show light brown wash between triangle (a prominent cell near the wing-base) and rear margin. Sexes similar (male's hindwing anal angle is rounded), but females have shorter, leaf-shaped abdominal appendages.

Habitat & Habits Found at still waterbodies in open and disturbed habitats such as large drains, ponds, lakes and reservoirs; also landward edges of mangroves. Active and conspicuous insect that tirelessly patrols perimeters of ponds and other large waterbodies from late morning until late afternoon, dwarfing and overshadowing other dragonflies in its territory. Females active at breeding sites from late afternoon until about dusk, seeking out suitable substrates such as a half-submerged plant stem or underside of a lily pad. After spending a few minutes to insert some eggs, female flies off to search for another oviposition spot nearby. Oviposition while in tandem has also been observed.

Presence in Singapore Recorded in many locations with suitable habitats, including the Central Catchment Nature Reserve, Windsor Nature Park, the Singapore Botanic Gardens, Yishun Park and Pulau Ubin.

Etymology *Anax* is Greek for 'master' or 'ruler', possibly a reference to the insect's commanding presence in the air. Specific epithet, derived from *gutta*, Latin for 'drop', and *-atus*, 'marked with', refers to drop-shaped spots on sides of abdomen.

Distribution Tropical Asia, Japan, Australasia and Pacific Islands.

National Conservation Status Least Concern; Widespread but Uncommon.

IUCN Red List Status Least Concern.

Larva Typical of family, with large eyes, broad head and elongated body. Fierce and active hunter of aquatic prey such as tadpoles, small fish and insects, including other dragonfly larvae.

Dennis Farrell

Male patrolling a pond. Note the extensive blue on the base of abdominal segment 3. Photo taken in Thailand.

Tang Hung Bun

Pair in wheel.

Marcus Ng

Female inserting its eggs into a waterlily stalk. Note the slight brownish wash on the hindwing and the lack of a 'T'-shaped mark on the frons.

Robin Ngiam

Ovipositing female.

Robin Ngiam

The elongated larva.

ARROW EMPEROR *Anax panybeus* Hagen, 1867

Size HWL: c. 52 mm; TBL: c. 81mm

Description Very large green and blue dragonfly, very similar to the Emperor (p. 160). Can be distinguished from Emperor by dark 'T'-shaped mark on top of frons, though not very evident in some individuals, and abdominal segment 3, which is more constricted and elongated. Also less blue on abdominal segments 2 and 3 than in Emperor.

Habitat & Habits Found around ponds, lakes, slow streams and swamps. Probably overlooked due to similarity with Emperor and possibly more crepuscular habits. *Anax* species are strong flyers that can cover up to 100km in a day. This species' appearance in Singapore may represent a southwards dispersal from Peninsular Malaysia.

Presence in Singapore First Singapore record was at Pulau Ubin in 2017. Since then, has been sighted at various other locations, including Sungei Buloh Wetland Reserve, the Central Catchment Nature Reserve, Dairy Farm Nature Park, Jurong Eco-Garden and Pasir Ris Park.

Distribution South and Southeast Asia to Japan.

National Conservation Status Least Concern; Widespread but Rare.

IUCN Red List Status Least Concern.

Larva Similar to that of Emperor.

Michael Soh

Dorsal view of a male. Note the lack of blue on abdominal segment 3.

Lateral view of a male.

Female resting in a shady corner of Sungei Buloh Wetland Reserve.

SPOON-TAILED DUSKHAWKER *Gynacantha basiguttata* Selys, 1882

Size HWL: 46–49mm; TBL: 70–75mm

Description Large crepuscular dragonfly with dark brown streaks at wing-bases. Eyes and thorax green. Segments 1–2 of male's abdomen greatly swollen like a bulb, with bright blue and blue-green markings and semi-circular auricles. Segment 3 highly constricted and elongated. Abdomen ground colour black, with transverse green flecks. Upper appendages long and thin, with expanded, spatula-like tips that have dense tuft of hairs on inner side, which becomes denser near tip. Lower appendage almost half the length of uppers. Female similar (eyes may be brownish), but basal abdominal segments less inflated and segment 3 only slightly constricted. Distinguished from other local duskhawkers by larger size and dark streaks at wing-bases. *Gynacantha* is separated from the superficially similar nighthawkers (*Heliaeschna* species) by lack of crossveins in median space of both wings.

Habitat & Habits Found in swampy areas in forests. Breeds in leafy bottomed pools guarded by one or, if the pool is large enough, more males. After mating, female oviposits in leaf litter. Duskhawkers are crepuscular dragonflies that are active at dawn and dusk, hanging inconspicuously from low vegetation by day. This may be why they are under recorded. Occasionally attracted to artificial lights in buildings.

Presence in Singapore Recorded in the Bukit Timah and Central Catchment Nature Reserves, Mandai and Kent Ridge. In 2019, a male was sighted at Khoo Teck Puat Hospital, to which it may have strayed from the forested parts of the nearby Yishun Park.

Etymology Generic epithet combines the Greek for 'woman' (*gynaikos*) and 'thorn' (*akantha*). It refers to two ventral spines on female's last abdominal segment, which are used to pierce holes in plant matter or soil for oviposition.

Distribution Southeast Asia.

National Conservation Status Vulnerable; Restricted and Rare.

IUCN Red List Status Least Concern.

Larva Typical of genus, with slender and elongated body, large head and eyes, and rounded postocular lobes.

Cheong Loong Fah

Male in Bukit Timah, showing the darkened wing-bases that distinguish this species from other local duskhawkers.

Cheong Loong Fah

Female, also with darkened wing-bases.

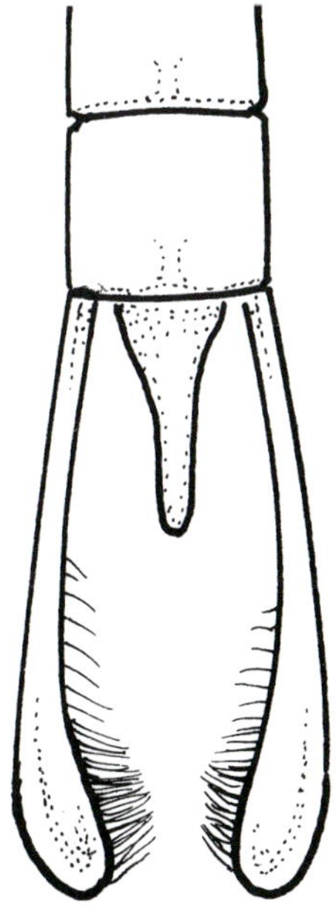

Albert G. Orr

Drawing of the male's anal appendages, showing the spatulate upper appendages.

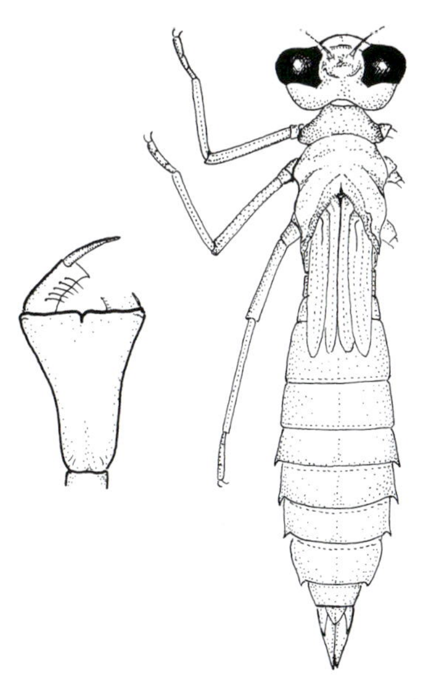

Albert G. Orr

Drawing of a Gynacantha species larva. The labial palp (shown on the left) has a row of setae (fine bristles), which is absent in other aeshnid genera.

Small Duskhawker *Gynacantha bayadera* Selys, 1891

Size HWL: 40–44mm; TBL: 56–60mm

Description Large crepuscular dragonfly, smaller than other local duskhawkers. Eyes green to blue. No 'T'-shaped mark on frons, unlike in the Spear-tailed and Dingy Duskhawkers (pp. 168 and 170). Wings hyaline. Male has light green thorax that is lighter on ventral side. Abdominal segments 1–2 inflated, but much less so than in other duskhawkers, with less rounded, more triangular, auricles. Segment 3 constricted. Upper appendages long and straight, not inflated at tips. Female similar to male but basal abdominal segments only slightly expanded.

Habitat & Habits Found in swampy forested areas. By day, rests amid dense undergrowth or in small gullies, often in inconspicuous positions not readily visible from trails. May sometimes be flushed from its hiding place and dart about, and even hover before an observer, before vanishing into another hidden perch. Such crepuscular habits may be an adaptation to take advantage of large swarms of small flies and micromoths that emerge at dusk. Duskhawkers also tend to cease their flight once insectivorous bats start to become active.

Presence in Singapore Recorded in several locations, including the Central Catchment Nature Reserve, Rifle Range forest, Labrador Nature Reserve, Queenstown and Pulau Semakau. One male was attracted to lights at the Nanyang Technological University campus in Jurong.

Etymology Specific epithet probably derived from *bailadeira*, a Portuguese term for 'female dancer'.

Distribution Tropical Asia from India to South China and New Guinea.

National Conservation Status Least Concern; Widespread but Rare.

IUCN Red List Status Least Concern.

Larva Unknown, but probably typical of genus.

Leonard Tan

The male can be told apart from other duskhawkers by its smaller size and abdomen-base, which is only slightly inflated.

Marcel Finlay

Mature male with blue eyes, showing the light green thorax.

Cheong Loong Fah

Older male.

Marcus Ng

Young male with undeveloped colours.

Cheong Loong Fah

A female.

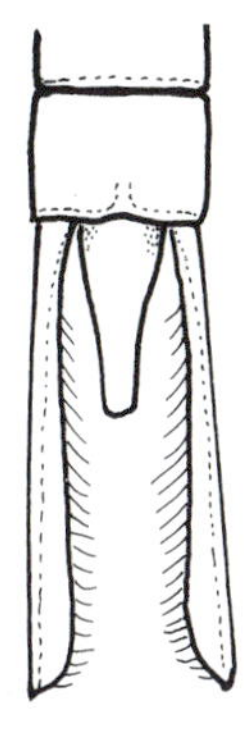
Albert G. Orr

Drawing of the male's anal appendages, showing the fairly straight upper appendages.

SPEAR-TAILED DUSKHAWKER *Gynacantha dohrni* Krüger, 1899

Size HWL: 42–44mm; TBL: 62–66mm

Description Large green crepuscular dragonfly with hyaline wings. Male has light green eyes and thorax. 'T'-shaped mark on frons, but not as well defined compared to that in the Dingy Duskhawker (p. 170). Abdominal segments 1–2 greatly inflated like a bulb, with bright blue auricles. Upper appendages have spear-shaped tips; hairs on inner side do not reach tip. Lower appendage pale and short, less than a third the length of uppers. Female similar but with duller colours. Appendages of female duskhawkers are often broken off or damaged, probably during mating or oviposition.

Habitat & Habits Found in swampy forests, both inland and near the coast. Hangs from low, often dense vegetation by day. Males guard shallow, leafy-bottomed pools in shaded forests. Females that arrive to inspect the site are seized. After mating, female returns to the pool to oviposit in leaf litter. Towards dusk, large numbers of duskhawkers may show up at forest clearings and canopies, hawking for small insects at high speeds.

Presence in Singapore Locally, first recorded in Tampines in 1979 (wrongly identified as *Gynacantha bayadera*). Now known from many locations, including the Central Catchment Nature Reserve, Hindhede, Windsor and Thomson Nature Parks, Sungei Buloh Wetland Reserve, Kranji, Pasir Ris, Pulau Semakau and Pulau Ubin.

Etymology Specific epithet honours Heinrich Wolfgang Ludwig Dohrn (1838–1913), a German entomologist, who provided Krüger with many dragonfly specimens from Sumatra.

Distribution Sundaland and the Philippines.

National Conservation Status Least Concern; Widespread but Uncommon.

IUCN Red List Status Least Concern.

Larva Unknown, but expected to be typical of genus.

Marcus Ng

Dorsal view of a male. Note the hyaline wings, greatly expanded basal abdominal segments and pale lower anal appendage.

Marcus Ng

Lateral view of a male. Males cling to vegetation in this position during the day, usually in low vegetation by swampy pools.

Tang Hung Bun

The male's pointed upper appendages and pale lower appendage. Note the inner rows of hair (with a water droplet in between), which end well before the tip.

Marcus Ng

Female duskhawker, probably Gynacantha dohrni, by a shaded stream in swampy forest.

Flora Li

Female ovipositing in leaf litter by a small shallow forest pool in Windsor Nature Park.

Marcel Finlay

Mating pair in Windsor Nature Park, where the species is often seen.

Dingy Duskhawker *Gynacantha subinterrupta* Rambur, 1842

Size HWL: 44–47mm; TBL: 67–72mm

Description Large crepuscular dragonfly with light blue eyes and hyaline wings. Upper surface of frons has well-defined, dark 'T'-shaped mark that is most visible when viewed dorsally. Male has green thorax. Abdominal segments 1–2 inflated like a bulb, with semi-circular auricles. Segment 3 strongly constricted. Remaining segments slender and dark, with blue-green markings. Upper appendages long and fairly straight, with hairs along inner edge. Lower appendage short and dark – contrast with pale lower appendage of the Spear-tailed Duskhawker (p. 168). Female similar but paler. Females of Dingy and Spear-tailed Duskhawkers very difficult to tell apart, as eye colours may be variable.

Habitat & Habits Found in swampy, forested areas; also densely wooded areas in parkland. By day, both sexes perch in dense masses of fallen branches and other low vegetation. May stray into urban areas, especially at night. Most widespread and common member of the genus locally. Fraser (1936) provided this observation of the genus, which still rings true: 'With rare exceptions all species are crepuscular by nature, not appearing on the wing until dusk has well set in. Their principal food appears to be mosquitos and microlepidoptera. During the day they may be flushed from dark thickets, especially bamboo in swampy low-lying country. When so flushed they soon find a new resting place, but one where it is impossible to take them with a net on account of the dense and thorny nature of the jungle.'

Marcus Ng

Male at Thomson Nature Park. Note the bluish eyes, 'T'-shaped mark on the frons and dark lower anal appendage.

Presence in Singapore Recorded in many locations with suitable habitats, including the Central Catchment Nature Reserve, Thomson Nature Park, Kent Ridge, Admiralty Park, Yishun Park, Kranji, Pasir Ris, the Singapore Botanic Gardens, Pulau Semakau and Pulau Ubin.

Distribution Widespread in tropical Asia.

National Conservation Status Least Concern; Widespread but Uncommon.

IUCN Red List Status Least Concern.

Larva General appearance typical of genus.

Marcus Ng

Female with intact anal appendages. The long cerci are often broken off after mating.

Robin Ngiam

Late instar larva.

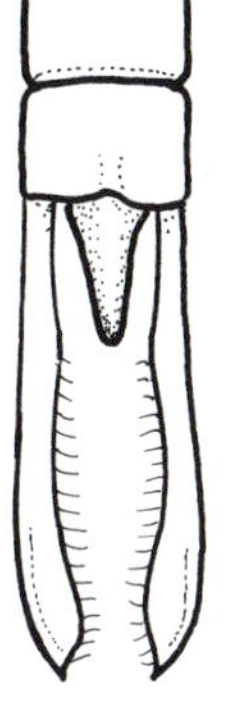

Albert G. Orr

Drawing of the male's anal appendages.

NIGHTHAWKER *Heliaeschna crassa* Krüger, 1899

Size HWL: 49–52mm; TBL: 72–77mm

Description Very large crepuscular dragonfly with 4–5 crossveins in median space of both wings, contrasting with genus *Gynacantha*, in which median space lacks crossveins. Male's eyes and thorax light green. Top of frons black, without 'T'-shaped mark. Wings hyaline, with slightly darkened bases. Abdominal segments 1–2 greenish and inflated like a bulb, with backwards pointing, triangular auricles. Segment 3 constricted. Rest of abdomen dark reddish-brown, with pale greenish transverse flecks on segments 3–8. Upper appendages long and curved slightly inwards. Lower appendage very short and slightly recurved upwards when seen in profile. Female brownish with olive-green to brownish eyes. Wings smoky brown and may have dark streak along costa. Very similar to *H. idae*, which is also widespread in Sundaland, and can be distinguished only by close inspection of male's appendages; females very difficult to tell apart. Some photographs of females taken in Singapore may prove to be *H. idae*. However, this genus requires more taxonomic work and the two species may in fact be a single species.

Habitat & Habits Occurs in swampy forests. Rests on vegetation during the day, typically becoming active around and well after sunset. Courtship and mating take place after dusk. May also be active during overcast weather or the late afternoon.

Presence in Singapore Recorded confidently in the Central Catchment Nature Reserve (MacRitchie forest). Records of females from Kent Ridge, Kranji and the National University of Singapore's Bukit Timah Campus may be this species or *H. idae*.

Etymology Generic epithet, combining a Greek term for 'forked tendrils' and *-aeschna*, a common appellation in hawker

Chien Lee

Male from Sarawak, Malaysia, showing the numerous crossveins in the median space of both wings.

names, may refer to a 'prominent forked plate' under abdominal segment 10 of female. Specific epithet may come from *crassus*, Latin for 'thick' or 'stout'. The other putative species, *idae*, is named after Ida Laura Pfeiffer (1797–1858), an Austrian explorer who first collected the dragonfly in Borneo.

Distribution Sundaland and South Thailand.

National Conservation Status Critically Endangered; Restricted and Very Rare.

IUCN Red List Status Least Concern.

Larva Unknown. Larvae of this genus are typical of aeshnids, with a large head, large eyes, dark brown body and banded legs.

Tang Hung Bun

Female Heliaeschna photographed in Singapore. Females of crassa and idae are difficult to tell apart and may in fact be conspecific.

Robin Ngiam

Larva of Heliaeschna idae photographed in Sarawak, Malaysia.

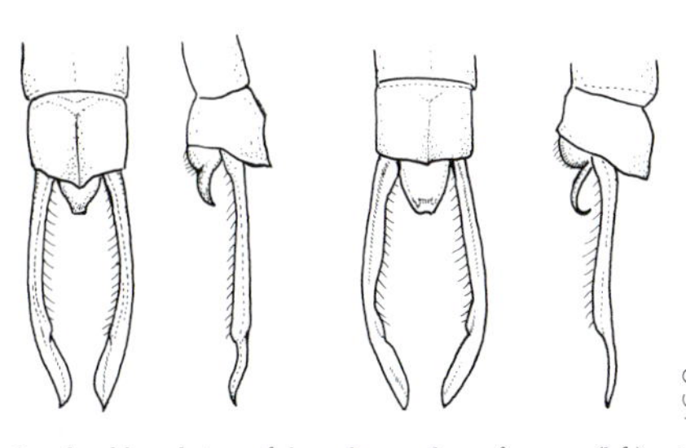

Albert G. Orr

Dorsal and lateral views of the male appendages of crassa (left) and idae (right).

PLAIN NIGHTHAWKER *Heliaeschna simplicia* (Karsch, 1891)

Size HWL: 45–46mm; TBL: c. 62mm

Description Large green dragonfly with five crossveins in median space of forewings. Wings mainly hyaline. Eyes bluish-green. Thorax green; legs brownish. Abdominal segments 1–2 inflated with green markings. Abdomen dark, marked by thin greenish flecks. Male's anal appendages broad with outwardly pointed tip and distinctive inner notch. Female similar but with thicker abdomen and duller colours.

Habitat & Habits Found in swampy forests. Crepuscular and elusive, which is typical of the genus. May be attracted to artificial lights.

Presence in Singapore New record for Singapore in September 2021 from Kranji Marshes, thus far the only known locality.

Etymology Common name 'plain' is coined from specific epithet, which is derived from *simplex*, meaning 'simple' or 'plain' in Latin.

Distribution Sundaland and the Philippines, north to Cambodia.

National Conservation Status Critically Endangered; Restricted and Very Rare.

IUCN Red List Status Least Concern.

Larva Typical of genus. Inhabits bottoms of swampy pools among mud and leaf litter. Can be differentiated from other *Heliaeschna* species larvae by more elevated, rounded head and inner structures of labium.

Veronica Foo Tse Fen

Dorsal view of a male at Kranji Marshes. The median space of the forewings has five crossveins.

Veronica Foo Tse Fen

Lateral view of the male that was discovered at Kranji Marshes in 2021.

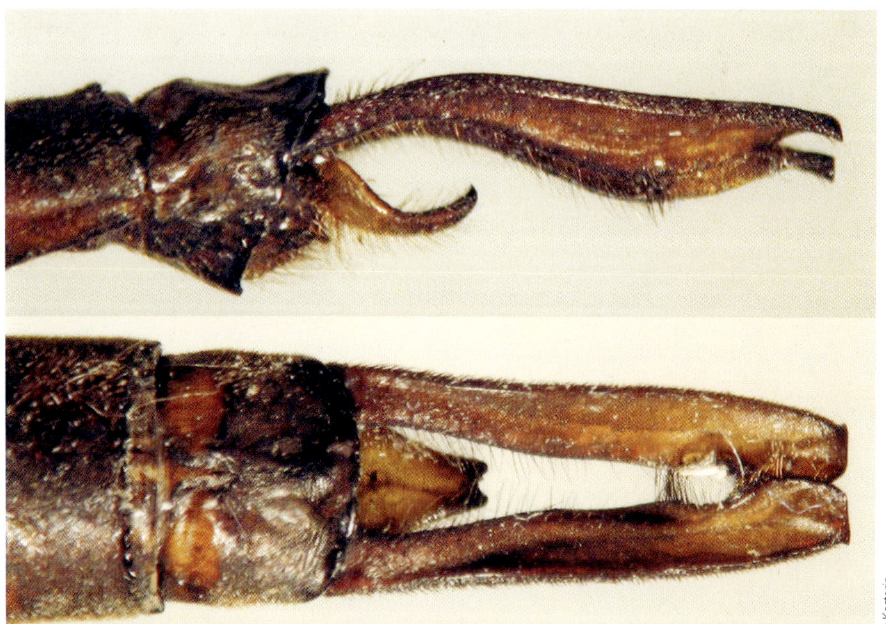

Oleg Kosterin

Lateral and dorsal views of the male's anal appendages. Note the notched lower appendage (dorsal view), which helps to distinguish this species from other nighthawkers.

LESSER NIGHTHAWKER *Heliaeschna uninervulata* Martin, 1909

Size HWL: 40–44mm; TBL: 64–69mm

Description Large dragonfly with single crossvein in median space of both wings (contrast with *Gynacantha* genus, members of which have no crossveins, and other nighthawkers, which have multiple crossveins). Sexes similar. Eyes greenish in male, greenish-yellow in female. Thorax olive-brown, legs brownish. Abdominal segments 1–2 slightly inflated and marked with green stripes; remaining segments brownish with dark bluish to greenish flecks and patches. Male's appendages distinctive, with uppers flat and broad like a leaf blade.

Habitat & Habits Found in swampy forests. Like other *Heliaeschna* species, active around dusk, feeding and mating even after dark. However, female has been seen to oviposit during the late afternoon, inserting eggs into small twigs overhanging a small, stagnant pool filled with leaf litter. May be attracted to artificial lights.

Presence in Singapore Recorded in the Central Catchment Nature Reserve, Windsor Nature Park, Mandai, Eng Neo forest, Pasir Ris and Pulau Ubin.

Etymology Specific epithet references single crossvein in median space of wings.

Distribution Southeast Asia.

National Conservation Status Near Threatened; Widespread but Rare.

IUCN Red List Status Least Concern.

Larva Typical of genus. Found in muddy forest pools filled with leaf litter. Ferocious hunter, often stalking and ambushing prey by crawling along submerged twigs before grasping its victim from above.

Tan Hui Zhen

Dorsal view of a male. Note the single crossvein in the median space of the wings.

Tan Hui Zhen

Lateral view of a male.

Tang Hung Bun

A female ovipositing at a small stagnant pool in Windsor Nature Park in the late afternoon.

PADDLETAIL *Oligoaeschna amata* (Förster, 1903)

Size HWL: 35–37mm; TBL: 52–56mm

Description Fairly large dragonfly with dark green eyes and spindle-shaped abdomen. Wings hyaline, with pale amber tint. Wing venation much less dense than in other local hawkers. Male has green thorax with dark transverse markings. Abdomen dark with green flecks; segments 1–2 slightly inflated, but less so than in most *Gynacantha* species, segment 3 constricted. Upper appendages short and broad, paddle-like as the name suggests, and without stalk. Female's abdomen has tapering tip and long, spatulate appendages held at right angles to each other. Female's wings often have reddish-brown wash.

Habitat & Habits Found in forested swampy areas. Crepuscular species that hangs from low vegetation by day. Orr observed that in Borneo, *Oligoaeschna* species typically start to become active up to two hours before sunset, much earlier than *Gynacantha*, *Heliaeschna* and *Tetracanthagyna* species. Females may be active by day, searching for oviposition sites in deep forest.

Presence in Singapore The first local record of this species was a male collected by German explorer Andreas Fedor Jagor (1816–1900) in the late nineteenth century. In recent times, all individuals recorded have been females from a single location (outskirts of Nee Soon Swamp Forest) within the Central Catchment Nature Reserve.

Etymology *Oligo* means 'few' in Latin – possibly an allusion to the more open venation of this genus compared to most other hawkers. *Amata* is Latin for 'beloved'.

Distribution Singapore, Borneo, and Peninsular Malaysia (probable).

National Conservation Status Critically Endangered; Restricted and Very Rare.

IUCN Red List Status Data Deficient.

Larva Currently unknown for the genus.

Rory Dow

Lateral view of a male photographed in Sarawak, Malaysia.

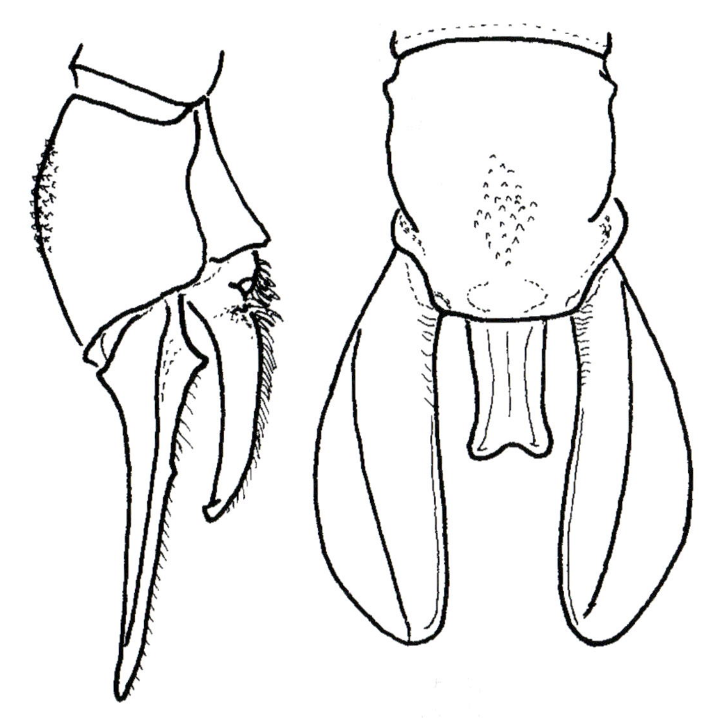

Albert G. Orr

Drawings showing lateral and dorsal views of the male's anal appendages, which are flattened when viewed from the side and also lack a petiole (stalk).

Michael Soh

Dorsal view of a female. Note the slight wash on the wings and bizarre anal appendages.

LEAFTAIL *Oligoaeschna foliacea* Lieftinck, 1968

Size HWL: 38–41mm; TBL: 59–62mm

Description Large dragonfly with dark green eyes (brownish with age) and unique appendages. Very similar to the Paddletail (p. 178), but male's upper appendages are leaf-like, with visibly stalked base. Female has racquet-like appendages, which are often lost, usually after mating. Wings may have reddish-brownish wash.

Habitat & Habits Occurs in wooded swampy areas. Active towards dusk, resting in dense undergrowth by day. Orr wrote of this species in Brunei: 'Both sexes hawked for insects along a disused logging road from between 1600 and sunset, at about 1800. Typically they flew about 5 m above the ground for the first hour of activity, descending to 1–2 m near dusk ... the conspicuous broadly spatulate appendages of the female are held well out from the body, suggesting a signalling function.' May be attracted to lights at buildings. Local observations suggest some seasonality in appearance of adults (March–May, September–October).

Presence in Singapore Recorded in Nee Soon Swamp Forest.

Etymology Specific epithet comes from *foliaceus*, Latin for 'leaf-like', referring to abdominal appendages.

Distribution Singapore, Peninsular Malaysia and Borneo.

National Conservation Status Critically Endangered; Restricted and Very Rare.

IUCN Red List Status Near Threatened.

Larva Unknown.

Leonard Tan

Dorsal view of a male.

Dorsal view of a female.

Lateral view of a female.

Lateral view of a male.

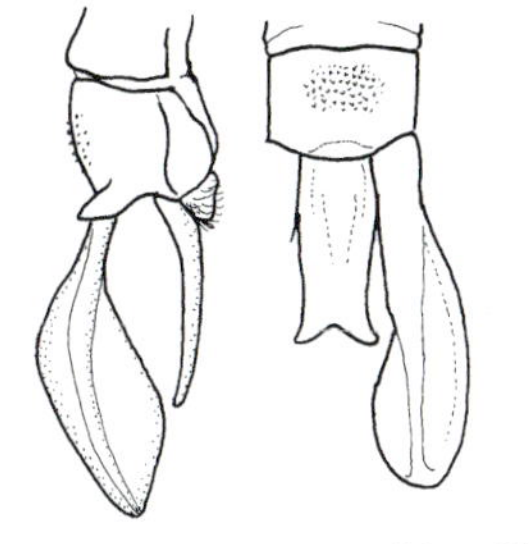

Drawings showing lateral and dorsal views of the male's anal appendages, which have a visible petiole (stalk).

Giant Hawker *Tetracanthagyna plagiata* (Waterhouse, 1877)

Size HWL: 67–76mm; TBL: 93–100mm

Description Largest dragonfly in region. Female is one of the world's largest living true dragonflies by wingspan and weight. Sexes similar but female larger than male. Eyes dark brown above, paler below. Thorax dark reddish or chocolate-brown, with broad, pale lateral bands. Abdomen reddish-brown without markings. Wings of both sexes have dark brown costal streak, but females may also have broad, transverse dark brown patches near wing-tips.

Habitat & Habits Found at forest streams and swampy forests. Forages at dawn or dusk high up in the canopy; perches on trees and shrubs by day. Females may be active throughout the day, seeking out dead branches and logs overhanging forest streams. When a suitable site has been chosen, female uses her ventral spines to probe and pierce the wood before inserting an egg, and repeats this action along the length of a branch.

Presence in Singapore Recorded in the Central Catchment Nature Reserve, Thomson Nature Park and Rifle Range forest.

Etymology Originally described as a species of *Gynacantha*. Sélys then assigned it to a new genus combining the Latin for 'four' (*tetr-*) and inverting the earlier generic elements *gyna* and *acantha*. Generic epithet refers to four ventral spines on last abdominal segment of female. *Plagiata* is Greek for 'marked obliquely', perhaps referring to slanted thoracic markings.

Distribution Sundaland and Southern Thailand.

National Conservation Status Vulnerable; Restricted and Uncommon.

IUCN Red List Status Least Concern.

Larva Found quite regularly during aquatic sampling, unlike the rather elusive adult. Impressive creature with elongated body and distinctive horn-like protuberance on head. Dwells amid leaf litter in clear, slow-flowing streams. Nocturnal and semi-terrestrial, with older larvae emerging at night to cling to a twig just above the water, from which they ambush small fish, tadpoles or shrimps close to the surface. Faeces are projected forcefully out of water. When threatened, larva exhibits thanatosis (feigning death).

Robin Ngiam

Tattered male in Nee Soon Swamp Forest. Note the dark costal streak.

Cheong Loong Fah

Female in Nee Soon Swamp Forest. Note the dark wing markings near the tips and dark costal streak.

Chris Ang

Female by a forest trail.

Robin Ngiam

Larva on a twig above the water, with a captured shrimp.

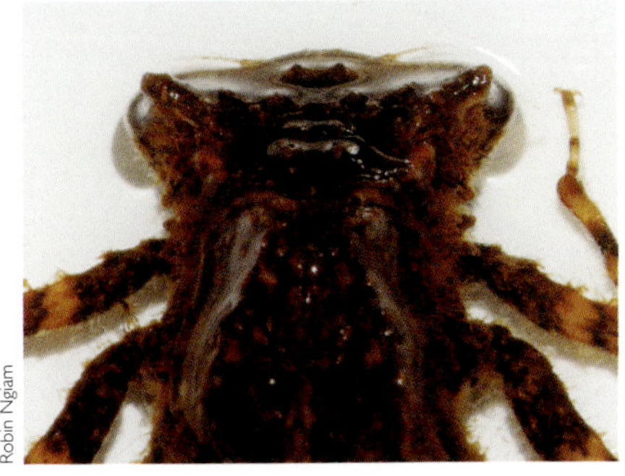

Robin Ngiam

Close-up of the larva's head, showing horn-like protuberances.

Robin Ngiam

Larva on a twig just above the water, holding a captured fish.

Corduliidae (Emeralds)

The emeralds are small to medium-sized, often metallic green or brown dragonflies with brilliant green eyes. The eyes of the male meet closely at the top of the head. The wings are mostly hyaline, with a well-developed anal loop. Emeralds are usually highly aerial in habit, coursing over streams and tree tops, often hovering. They perch in a vertical position like hawkers and cruisers.

With more than 150 species, this family is better known in the northern temperate zone, where there are many representatives in genera such as *Cordulia*, *Somatochlora* (the most northerly dragonfly genus) and *Epitheca* (baskettails), which breed in bogs, fens and the tundra.

The genus *Hemicordulia* is widespread in the Old World tropics, Australasia and the Pacific Islands, with more than 35 species. The family name is derived from the Holarctic genus *Cordulia*, in which the male abdomen is shaped like a club or cudgel (*kordyle* in Greek).

In the earlier book by Tang et al, the family Corduliidae included dragonflies in the genera *Epophthalmia, Macromia* and *Idionyx*. However, following taxonomic revisions in the past decade, *Epophthalmia* and *Macromia* have been placed in their own family, Macromiidae (p. 325), while *Idionyx* is now a putative member of the Synthemistidae (p. 332).

Damian Pinguey.

A male Cordulia aenea (Linnaeus, 1758), or Downy Emerald, in the UK, showing its cudgel-shaped abdomen and the metallic green colours typical of the family.

EMERALD *Hemicordulia tenera* Lieftinck, 1930

Size HWL: 28–29mm; TBL: 45–47mm

Description Medium-sized, lightly built green dragonfly. Eyes light green. Body dull metallic green, with slender abdomen. Appendages long and slender, especially uppers, which are curved slightly inwards when viewed from above. Bases of hindwings rounded in both sexes. Female similar to male, with moderately long cerci.

Habitat & Habits Found around marshes, small lakes and streams in forests and forest edges. A rare species, usually recorded in higher altitudes elsewhere in region. Known to hawk actively over the forest canopy, hilltops and ridges, where it is most frequently seen. May be attracted to artificial lights at night.

Presence in Singapore Recorded in various locations in recent years, including Kent Ridge (attracted by lights at the university campus), the summit of Bukit Timah, Sentosa (Mount Serapong), Sungei Buloh Wetland Reserve and Telok Blangah Hill Park.

Etymology *Hemi* means 'half' in Greek. *Hemicordulia* was originally established as a subgenus of *Cordulia* but later raised to a genus of its own. Specific epithet may be derived from *tener*, Latin for 'delicate' or 'tender'.

Distribution Sundaland, Thailand, and Myanmar (probable).

National Conservation Status Vulnerable; Restricted and Rare.

IUCN Red List Status Least Concern.

Larva Unknown, but expected to be similar to known larvae in genus, which are squat, with ovoid abdomen and long legs. Probably breeds in marshes and wetlands in forests and forest edges.

Lee Yan Leong

Dorsal view of a female found on the twenty-sixth storey of a building in Tiong Bahru in 2009.

Lee Yan Leong

Lateral view of the same female.

Rory Dow

Male photographed in Sarawak, Malaysia.

Martin Kennewell

Male hawking for prey high above the summit of Bukit Timah Hill.

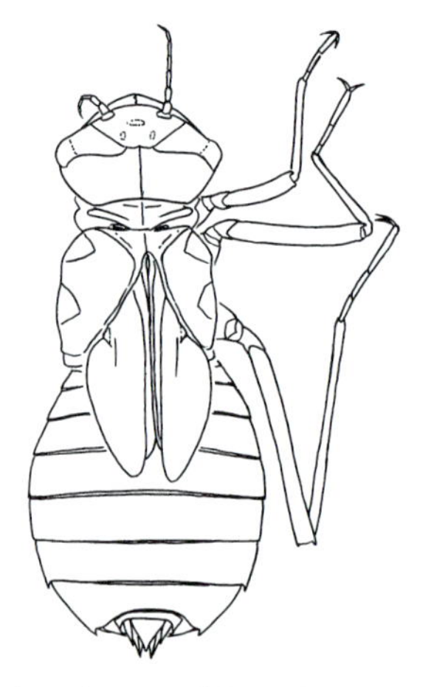

Gunther Theischinger

Drawing of the larva of Hemicordulia tau *from Australia, showing the typical features of the genus.*

Robin Ngiam

Dorsal view of the male's anal appendages.

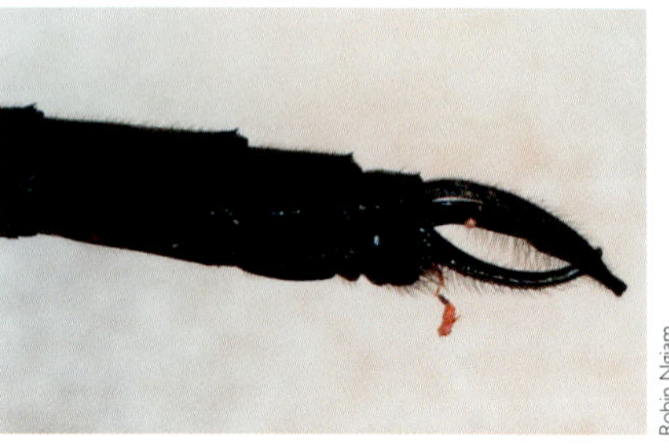

Robin Ngiam

Lateral view of the male's appendages.

Gomphidae (Clubtails)

The family Gomphidae comprises small to very large dragonflies, many of which have a slender abdomen with swollen, club-like terminal segments, hence the name 'clubtail'. The head is relatively small, with widely separated, usually greenish eyes. The legs are relatively short, and local species usually perch horizontally on a leaf, twig or rock, never in a hanging position like the long-legged hawkers and cruisers. Males and some females have small auricles on the sides of the second abdominal segment. The wings are long, hyaline and fairly narrow, with an acute anal angle in the male hindwing-base. The flight of clubtails is strong, swift and commanding, and on sunny days forest-dwelling clubtails may be seen hunting and chasing each other over the canopy at great speeds. The 'club' is thought to serve as a counterbalance during tight aerial manoeuvres.

Apart from the Common Flangetail, all local clubtails are forest dependent and breed in clear, unpolluted streams. They are seldom encountered, even by keen dragonfly watchers, due to their fugitive and unpredictable habits; males may come down to breeding sites for just an hour or so each day, while females may linger for only about 15 minutes. Writing of the family in Borneo, Orr (2003) remarked: 'They are mainly active around the middle of the day. Many species spend the bulk of their time in the forest canopy and approach the water only briefly to mate and oviposit.' In contrast to the adults, the larvae of some forest-dwelling species can be quite abundant during sampling.

Clubtails are particularly susceptible to water pollution and environmental disturbance. Some species, such as Ris' Clubtail and the Banded Hooktail, are dependent on streams with small, level sand bars, which they require for emergence. Worldwide, there are more than 1,000 species, with 11 known from Singapore. The family name is derived from the Palearctic genus *Gomphus*, which was named for its clavate (club-shaped) abdomen. In Greek, *gomphos* refers to a bolt or rivet used in shipbuilding. Many clubtails have the suffix *-gomphus* as part of their generic epithets.

Erland Refling Nielsen

Male Gomphus vulgatissimus *(Linnaeus, 1758), or Common Clubtail, in Sweden. The type species for the entire family, this clubtail is found at rivers and creeks across Europe.*

Malayan Hooktail *Acrogomphus malayanus* Laidlaw, 1925

Size HWL: 29–33mm; TBL: 43–45mm

Description Small-medium dragonfly with short legs and distinctive markings. Synthorax has two thin yellow stripes on dorsum and two thicker, well-separated lateral stripes. Basal part of each abdominal segment marked with yellow except segments 9–10, which are black. Yellow markings may have greenish tinge in older individuals. Sexes similar.

Habitat & Habits Forest-dependent species that breeds in streams, where larvae burrow under sand. Adults mainly canopy dwelling. Females may be seen ovipositing at streams using backwards and forwards flight pattern. Little is known about habits, but Fraser wrote of closely related Indian species: 'Species of the genus are arboreal by nature, often resting at great heights on trees, only occasionally coming down to the beds of turbulent mountain streams, in which they breed in wild areas.' He noted their 'habit of roosting at great heights, often as much as 100ft (about 30.5m) or more above the ground', adding that 'with the aid of field-glasses males could be seen at times perched on prominent dead twigs on the tops of trees, or soaring at great heights'.

Presence in Singapore The first local records were two males collected in 2012 from malaise traps, followed by larvae found and reared by Robin Ngiam. Larvae can be quite numerous at known locations, so it is surprising that the species was not reported by earlier researchers. Recorded (mostly larvae) in Nee Soon Swamp Forest, Thomson and Windsor Nature Parks and Rifle Range forest.

Etymology In Greek, *acro-* means 'pointed' or 'high'.

Distribution Endemic to Singapore and Peninsular Malaysia. Known from only a dozen or so locations.

National Conservation Status Vulnerable; Restricted and Rare.

IUCN Red List Status Near Threatened.

Larva Found in forest streams with sandy or gravelly bottoms. Body torpedo shaped, with very bulging, seemingly muscular prothorax. Wing-buds divergent. Very active burrower with strong legs, capable of disappearing into the substrate in a matter of seconds, expelling squirts of water from the anus while digging. Voracious predator.

Robin Ngiam

The torpedo-shaped larva, showing the divergent wing-buds.

Kenneth Fong

Young male from Windsor Nature Park. The appendages are much smaller compared to those of the male Banded Hooktail.

Erland Refling Nielsen

Dorsal view of a female photographed in Malaysia. Note the pale abdominal markings, which are smaller and less extensive compared to those of the female Banded Hooktail.

Arthur's Clubtail *Burmagomphus arthuri* Lieftinck, 1953

Size HWL: c. 24mm; TBL: c. 35mm

Description Small, pale greenish clubtail with short legs. Eyes bluish-green. Synthorax brownish with two pale green dorsal stripes, laterally mostly olive-green. Wings hyaline. Abdomen segments 1–4 greenish-brown, thereafter becoming mainly brownish with faint greenish-yellow markings. Female easily recognized by strongly downwards curving vulvar scale (pseudo-ovipositor) beneath segment 8. Originally described from female from Borneo. Male remains undescribed, but known from specimens collected in Borneo by Rory Dow.

Habitat & Habits Ecology unknown but probably typical of elusive, mainly arboreal forest gomphids. May be encountered perching in sunlit spots on foliage beside streams, but very wary and readily flies straight up to the canopy at the slightest disturbance. Locally, females inhabit small, slow-flowing forest streams with sandy or mud bottoms, where they are usually seen flying, ovipositing in shady stretches under foliage.

Presence in Singapore First recorded (females) in Singapore in 2012. A female larva was reared successfully in 2014. Known only from Nee Soon Swamp Forest.

Etymology The genus, which consists of rather small, gracile clubtails, was coined by E. B. Williamson in 1907, based on a species discovered in Burma (now Myanmar). Species named after Arthur M. R. Wegner (1894–1969), an entomologist active in Indonesia in the 1930s–'60s, who discovered the female holotype.

Distribution Singapore, Peninsular Malaysia, southern Thailand and Borneo, with very few records.

National Conservation Status Critically Endangered; Restricted and Very Rare.

IUCN Red List Status Data Deficient.

Larva Small and flat. Early instars have triangular-shaped abdomen that becomes more elongated in later stages. Parallel wing-buds. Slow-moving and lives buried among substrate of sand or mud and leaf litter.

Robin Ngiam

Lateral view of a female, showing the distinctive long pseudo-ovipositor.

Robin Ngiam

Early instar larva, showing its flat body and triangular-shaped abdomen.

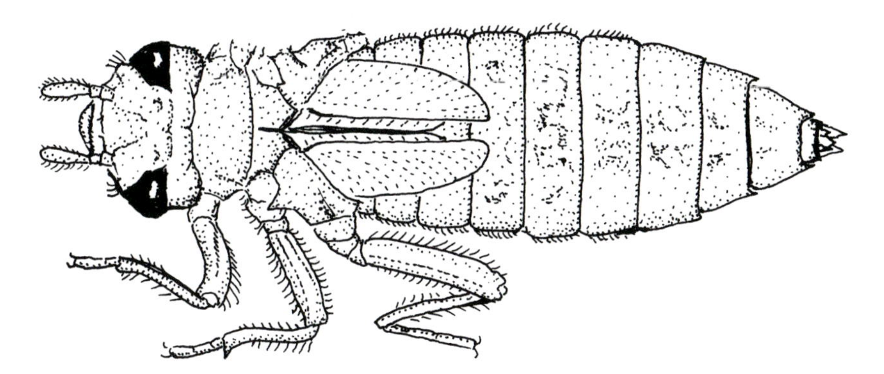

Drawing of the mature larva of Burmagomphus williamsoni, *a closely related species found in mainland Southeast* Asia.

SPLAYED CLUBTAIL *Burmagomphus divaricatus* Lieftinck, 1964

Size HWL: 22–23mm; TBL: 37–38mm

Description Small, black and green gomphid with green eyes and distinctive yellowish-green thoracic pattern. Dorsum of synthorax bears two green stripes that form shallow, 'V'-shaped green marking and two shorter slanted green stripes in front of 'V'. Sides of synthorax have green and black banding that may vary in width between individuals. Abdomen slender and predominantly black with small green markings. Appendages short; uppers with tips curved slightly outwards; lower appendage has strongly divergent arms. Female similar to male, but may appear more yellowish.

Habitat & Habits Found around small, sluggish muddy streams in somewhat open, disturbed forests. Arboreal by nature.

Presence in Singapore Locally, only record is of larva collected by D. S. Johnson in 1956 from the upper reaches of Sungei Seletar in Nee Soon Swamp Forest.

Etymology In Latin, *divaricatus* means 'spread apart', probably referring to strongly divergent male lower appendage.

Distribution Mainland Southeast Asia to China (Yunnan).

National Conservation Status Extinct.

IUCN Red List Status Least Concern.

Larva Typical of genus in general appearance.

Dennis Farrell

Dorsolateral view of a male photographed in Thailand, showing the 'V'-shaped marking on the synthorax dorsum.

Albert G. Orr

Dorsal view of the synthorax.

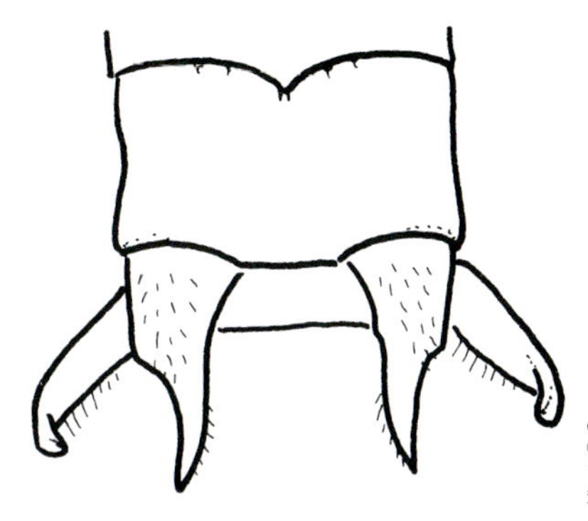
Albert G. Orr

Dorsal view of the male's anal appendages.

Dennis Farrell

Female photographed in Thailand, with similar markings to the male.

Matti Hämäläinen

Female photographed in Thailand, extruding a batch of eggs from her abdomen.

Dennis Farrell

Pair in wheel photographed in Thailand.

LESSER SPLAYED CLUBTAIL *Burmagomphus plagiatus* Lieftinck, 1964

Size HWL: 21–23mm; TBL: 36–38mm

Description Small greenish gomphid. Very similar to the Splayed Clubtail (p. 192), but synthorax markings differ: dorsum has broader greenish stripe and lacks a 'V'-shaped marking. Side of synthorax has complete and fairly broad, yellowish-green median stripe. Upper abdominal appendages short and curved slightly inwards; lower appendage thicker and less divergent than in the other species. Sexes similar.

Habitat & Habits Found around slow- or fast-flowing streams with a muddy or sedimented substrate in forests or open habitats. Arboreal by nature.

Presence in Singapore Locally, the larva, doubtfully referred to as this species, was collected by D. S. Johnson in 1956 in Sungei Seletar, Nee Soon Swamp Forest. This remains the only Singapore record.

Distribution Singapore, Peninsular Malaysia, Borneo and Sumatra. Overall very few records.

National Conservation Status Extinct.

IUCN Red List Status Data Deficient.

Larva Typical of genus in general appearance.

Choong Chee Yen

Male photographed in Malaysia.

Albert G. Orr

Drawing of a male. Note the side of the synthorax, which has complete stripes, unlike that of the Splayed Clubtail, which has an incomplete black stripe.

Albert G. Orr

Dorsal view of the synthorax.

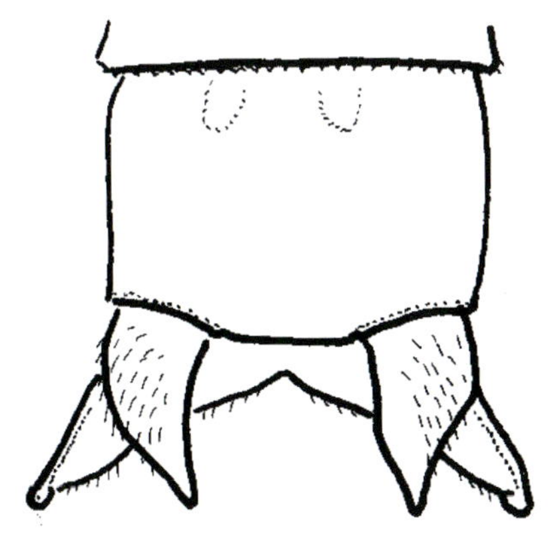

Albert G. Orr

Dorsal view of the male's anal appendages.

MALAYAN GRAPPLETAIL *Heliogomphus kelantanensis* (Laidlaw, 1902)

Size HWL: 25–26mm; TBL: c. 37mm

Description Smallish, slender gomphid with bluish-green eyes. Thorax olive-green on sides, with darker dorsum bearing pale green, 'L'-shaped markings. Hindwing has no discernible anal loop. Abdomen slender and mostly dark, becoming thicker from segment 7. Male appendages very short and dark. Sexes similar.

Habitat & Habits Elusive species of small, clear-flowing forest streams with ample amount of leaf litter. Arboreal dragonfly that favours high perches in the canopy. Adults may appear at sunlit spots along streams or nearby trails in the late morning, sometimes perching for long periods. Less wary than most other arboreal gomphids, preferring to fly to a nearby perch when disturbed rather than straight towards the canopy.

Presence in Singapore Recorded in Nee Soon Swamp Forest. Vagrant male was caught in a malaise trap in Pulau Ubin in 2012.

Etymology Helios is the Greek god of the sun. Laidlaw coined the genus for a group of clubtails from South and Southeast Asia, which could be regarded as 'sunny' regions, though the name also aptly describes their habits.

Distribution Singapore and Peninsular Malaysia. Currently known only from three locations in its range.

National Conservation Status Critically Endangered; Restricted and Rare.

IUCN Red List Status Vulnerable.

Larva Found amid sediment and leaf litter in fast-flowing water. Body broad, flat and ovoid, with well-expanded second antennal segment. Does not appear to burrow. Slow moving and well camouflaged among leaf litter.

Tang Hung Bun

Lateral view of a male, showing the 'L'-shaped dorsal markings and mostly green side of the synthorax.

Leonard Tan

Dorsal view of the male, showing the narrow wings and dark abdomen.

Marcel Finlay

Dorsal view showing the 'L'-shaped markings of the synthorax.

Marcel Finlay

Lateral view of a male.

Robin Ngiam

Larva, showing the highly inflated second antennal segment.

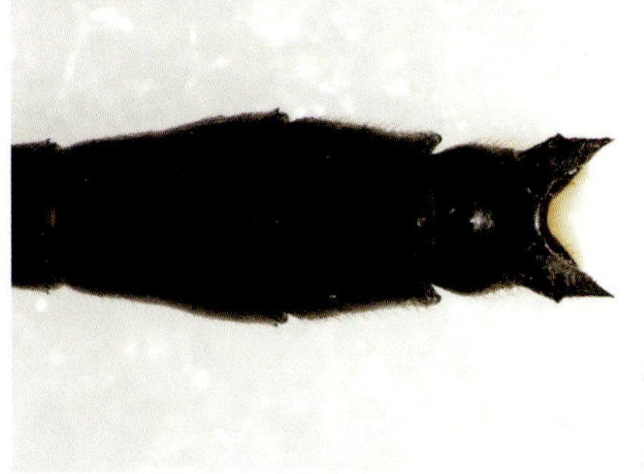

Tang Hung Bun

Close-up of the male appendages.

Common Flangetail *Ictinogomphus decoratus* (Selys, 1854)

Size HWL: 37–40mm; TBL: 64–68mm

Description Large, heavily built dragonfly with tiger-like stripes, short legs and prominent flanges on abdomen. Eyes dark greyish-green. Thorax and abdomen black with strong greenish-yellow stripes and streaks. Abdomen has pair of broad, rounded flaps on underside of segment 8. Appendages long and black. Female similar to male but has smaller flanges. Most common clubtail in Singapore. Easily distinguished from skimmers that share the same ponds by larger size, tiger-like patterns and abdomen shape.

Habitat & Habits Prefers lentic (still-water) habitats, but may also occur at small streams and swamps in open country and forest edges. Basks high in trees early in the day, coming down to the water later in the morning. Usually perched at tip of a leaf or twig in a horizontal or slanted posture, or with abdomen in semi-obelisk. Fast flying as males hunt or patrol their territory along the water's edge. Females less often seen by the water, usually foraging higher up or further away. Known to prey on smaller dragonflies; in turn often taken by bee-eaters (Meropidae).

Presence in Singapore Found across the island at large ponds, lakes and reservoirs.

Etymology *Ictinus* is Greek for 'kite', and may be a reference to large size of these dragonflies, or to abdominal flanges, which recall a kite's tail. Specific epithet means 'embellished' or 'decorated' in Latin.

Distribution Southeast Asia and south-west China.

National Conservation Status Least Concern; Widespread and Common.

IUCN Red List Status Least Concern.

Larva Dark, broad and flattened, with roundish abdomen. Inhabits bottom debris. Has been found on undersides of lily pads.

Marcus Ng

Lateral view of a male, showing the thoracic markings and prominent flange near the tip of the abdomen.

Marcus Ng

Frontal view of male photographed in Gopeng, Malaysia, showing the well-separated eyes.

Catalina Tong

Lateral view of a female, showing the reduced flange.

Catalina Tong

Dorsal view of a female, showing the rounded hindwing anal angle.

Marcus Ng

Common Flangetail feeding on a Common Parasol.

Robin Ngiam

The dark and flattened larva.

Ris' Clubtail *Leptogomphus risi*
Laidlaw, 1932

Size HWL: 29–30mm; TBL: c. 45mm

Description Medium-sized, green-yellow clubtail with slender abdomen and greenish eyes. Thorax black with two almost parallel, light greenish-yellow stripes on dorsum. Sides of thorax black with yellow bands and markings. Hindwing has no discernible anal loop. Abdomen dark with yellow auricles on segment 2. Upper abdominal appendages short and rounded, pale dorsally; lower appendage very broad. Sexes similar.

Habitat & Habits Occurs at forest streams. Adults spend most of the day in the canopy, coming down to breeding sites very briefly on sunny days. May sometimes be seen along forest trails or in clearings. Commenting on the genus, Orr (2003) wrote: 'Most members of this genus are rarely seen, and often it is the females that appear briefly as they come to drop their eggs in the water before flying back into the canopy.'

Presence in Singapore Rare and local in the Bukit Timah and Central Catchment Nature Reserves; also Dairy Farm Nature Park.

Etymology *Lepto* is Greek for 'thin' or 'delicate', and refers to the slender abdomen. Specific epithet honours Friedrich Ris (1867–1931), a Swiss doctor and entomologist who operated a psychiatric clinic in Rheinau, Switzerland.

Distribution Singapore, Peninsular Malaysia and southern Thailand.

National Conservation Status Vulnerable; Restricted and Rare.

IUCN Red List Status Least Concern.

Larva Flat, oviform and very hairy. Third antennal segment flat and elongated. Wing buds divergent. Found buried quite deep in a substrate of sand and leaf litter, usually in swifter flowing sections of streams.

Martin Kennewell

Dorsal view of a male at Wallace Trail, showing the yellow auricles, pale upper appendages and broad lower appendage.

Martin Kennewell

Lateral view of a male basking along Wallace Trail.

Phil Benstead

Female basking along Wallace Trail.

Cheong Loong Fah

Male showing the slender abdomen bearing pale auricles.

Robin Ngiam

Larva of the closely related Leptogomphus williamsoni photographed in Sarawak, Malayaia.

FORKTAIL *Macrogomphus quadratus* Selys, 1878

Size HWL: 44–48mm; TBL: 72–80mm

Description Huge, tiger-striped dragonfly with very long and slender abdomen. The largest clubtail in Singapore. Eyes dark green. Thorax and abdomen-base have black and pale yellow to greenish markings. Rest of abdomen black, with five orange-yellow bands. Upper appendages of male bifurcated, hence the name 'forktail'. Female similar to male but abdomen thicker and rear margin of hindwing rounded.

Habitat & Habits Found around slow-flowing streams in dense forests. Males patrol small, shallow streams, resting occasionally on twigs while vibrating their wings. Females lay their eggs in quick passes over shallow water in muddy banks. Both sexes may be seen basking on low branches along trails and in clearings from late morning until late afternoon. May occasionally stray into very open areas, such as car parks and grassy edges of nature reserves. While basking, flight may be fluttery and easy to follow, but when alarmed by a sudden movement, can take off and vanish in the blink of an eye.

Presence in Singapore Recorded in the Central Catchment and Bukit Timah Nature Reserves, and Old Upper Thomson Road.

Etymology *Makros* is Greek for 'long', and refers to the very long abdomens of species in this genus. Specific epithet means 'square' in Latin.

Distribution Singapore, Peninsular Malaysia, Borneo and Sumatra.

National Conservation Status Vulnerable; Restricted and Uncommon.

IUCN Red List Status Least Concern.

Larva Quite often found in aquatic surveys. Elongated and burrows into silty bottoms, using long abdominal 'snorkel' to access clean water. When threatened, may lift 'snorkel' like a scorpion's tail in a defensive display.

Marcus Ng

Dorsal view of a male showing the acute anal angle of the hindwing and bifurcated anal appendages.

Marcus Ng

Lateral view of a male basking at Golf Link.

Marcus Ng

Dorsal view of a female, showing the rounded hindwing anal angle and thicker abdomen.

Marcus Ng

Lateral view of the female.

Robin Ngiam

The elongated larva with its long rear 'snorkel'.

Malayan Spineleg *Merogomphus femoralis* Laidlaw, 1931

Size HWL: 26–27mm; TBL: c. 45mm

Description Small-medium greenish gomphid with very long and expanded hindlegs. Eyes green. Male's thorax has black and green stripes. Abdomen black with green-yellow banding. Appendages squat, with uppers pale and lower appendage dark and wide. Easily distinguished from other local clubtails by very long hindlegs, particularly the femora, which are very spiny and expanded. Female similar, but has thicker abdomen bearing white bands. Segment 1 of female's abdomen has greenish-white lines on dorsum and sides.

Habitat & Habits Found at acidic (low pH) streams in forests and swamp forests. Forages in the canopy, but comes down to perch low on leaves or logs at edges of streams.

Presence in Singapore Locally, restricted to Nee Soon Swamp Forest.

Etymology *Meros* is Greek for 'thigh' and refers to expanded hind femora of genus, as does specific epithet.

Distribution Singapore, Peninsular Malaysia and Borneo. Currently known only from Singapore and two sites in Sarawak. Likely extirpated from its type locality, Kuala Lumpur, where first specimen was collected in 1921.

National Conservation Status Critically Endangered; Restricted and Very Rare.

IUCN Red List Status Endangered.

Larva Elongated, with large head. Parallel wing-buds. Burrows shallowly in gravel or sand. Sluggish moving. Very similar to *Burmagomphus* species larva but distinguished by lack of burrowing hooks on legs and more elongated abdominal segment 9.

Tang Hung Bun

A male showing its very long and expanded hindlegs.

Veronica Foo Tse Fen

Female observed at the edge of swampy forest at Upper Seletar.

Robin Ngiam

Dorsal view of the male's anal appendages.

TINY SHEARTAIL *Microgomphus chelifer*
Selys, 1858

Size HWL: 19–20mm; TBL: 32–33mm

Description Very small clubtail – smaller than most local skimmers – with green eyes. Dorsum of synthorax has greenish-yellow, '7'-shaped markings; sides yellow with thin black bands. Hindwings very narrow, with no discernible anal loop. Abdomen black with small, greenish-yellow markings. Appendages black; uppers divided. Female similar to male. The *chelifer* species group requires taxonomic revision.

Habitat & Habits Found around shaded streams with sandy bottoms, in forests and forest edges. Males usually perch on leaves several metres above the water, occasionally coming down to land on sand bars and rocks. Most encounters are with ovipositing females. Fraser wrote of the genus in India: 'Species of the genus are arboreal by nature, but quite occasionally the males descend and settle on rocks in mid-stream; they do not appear to wander far from their parent streams, and may be found settled on evergreens, usually beside the water.'

Presence in Singapore Recorded in the Central Catchment and Bukit Timah Nature Reserves, also Dairy Farm Nature Park, Windsor Nature Park and Rifle Range forest.

Etymology *Mikros* is Greek for 'small', and describes this genus of clubtails. Specific epithet may combine the Greek for 'claw' (*chele*) and 'carry' (*fero*).

Distribution Southeast Asia, possibly north to Myanmar.

National Conservation Status Vulnerable; Restricted and Rare.

IUCN Red List Status Least Concern.

Larva Small, broad and flat. Quite similar in appearance to larvae of *Heliogomphus* species but antennae not expanded. Lives among substrate of leaf litter and sand.

Leslie Day

Dorsal view of a male in Koh Samui, Thailand, showing the distinctive dorsal markings and anal appendages.

Leslie Day

Lateral view of a male photographed in Koh Samui, Thailand, showing the distinctive dorsal markings and anal appendages.

Fiora Li

Female photographed in Windsor Nature Park.

Female extruding some eggs while resting on a leaf. She will then fly down to the water to release the eggs.

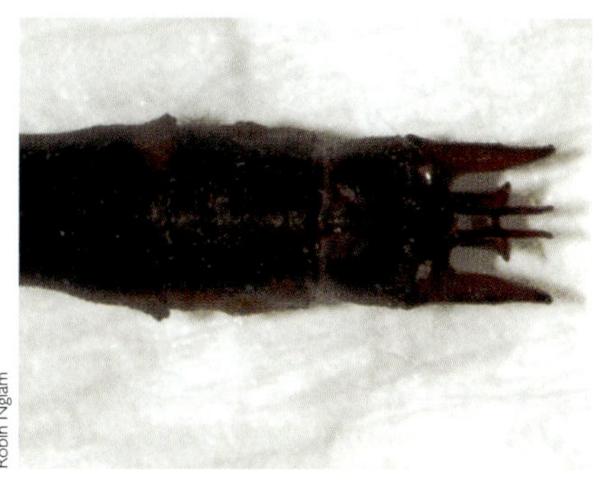

Robin Ngiam

Dorsal view of the male's appendages, showing the deeply divided upper appendages.

Robin Ngiam

The small and flat larva.

Banded Hooktail *Paragomphus capricornis* (Förster, 1914)

Size HWL: 22–24mm; TBL: 41–45mm

Description Small-medium, strongly built clubtail with extravagant male appendages. Eyes greyish-green. Thorax and abdomen black with bright orange-yellow bands and spots. Male's upper appendages very long and hooked downwards like scimitars. Female similar but without long appendages. Superficially, females may appear similar to the Malayan Hooktail (p. 188), but yellow patterns on abdomen are more extensive and broader, and abdomen-tip (segments 7–9) is less expanded.

Habitat & Habits Found around clear, shallow, low-gradient streams with sandy or gravelly bottoms in semi-open country near forests. Occasionally in denser forests if suitable streams exist. Towards midday, males come down to streams, resting on gravel beds or low rocks next to the water. When not at breeding sites, forages from the canopy of nearby forest trees.

Presence in Singapore Recorded in the Bukit Timah and Central Catchment Nature Reserves, and the forest off Mandai Road.

Etymology In Greek, *para* means 'beside' or 'close to'. Species was originally placed in the genus *Mesogomphus* ('middle gomphus') by Friedrich Förster, who saw similarities with *Gomphus* and *Onychogomphus* (another genus with extravagant anal appendages). Specific epithet means 'having goat-like horns'. Laidlaw remarked that 'the anal appendages of the male are shaped exactly like a tiny pair of chamois horns'.

Distribution Mainland Southeast Asia to southern China.

National Conservation Status Endangered; Restricted and Rare.

IUCN Red List Status Least Concern.

Larva General appearance very similar to larvae of *Acrogomphus*, with divergent wing-buds but without the greatly developed prothorax. Lives buried in sand.

Marcus Ng

Male photographed in Gopeng, Malaysia, showing the scimitar-like anal appendages.

Marcus Ng

Male photographed in Gopeng, Malaysia, perched by a shallow sandy stream flowing through open country near forest.

Leonard Tan

A female. The abdomen is marked as the male, but is stouter and lacks long anal appendages.

Wu Hongdao

The larva, showing divergent wing-buds. Photo taken in Guangdong, China.

Libellulidae (Skimmers)

The skimmers are a family of very small to fairly large, 'typical' dragonflies. Most of the common and conspicuous true dragonflies spotted in urban habitats, as well as wetlands and forests, belong to this family.

This is the largest dragonfly family, with about 1,040 species worldwide and 57 in Singapore. The eyes of skimmers are large and always contiguous to some degree. The wings are unstalked and may be hyaline, partially or fully coloured; the hindwings of both sexes have a rounded anal angle. Males lack auricles. The hindwing-base is often expanded and, in most larger species, features a sock-shaped anal loop. Females lack a functional ovipositor, but those of some species have a prominent vulvar scale or pseudo-ovipositor. Some females also have a pair of ventral flaps under segment 8; these help to hold the eggs before they are released into the water.

Many skimmers are perchers that hunt or guard territories at the margins of ponds, lakes, marshes and swampy pools. The family also includes highly aerial species that spend most of the day in the air, gliding or soaring with their greatly expanded hindwings

The males of many species are red, blue or a combination thereof (for example the grenadiers). The females are usually less colourful, but andromorphs are not uncommon in some species. In many species, older individuals of both sexes may develop a waxy pruinescence that obscures their underlying colours.

The family name comes from *Libellula*, a name used by Linnaeus to sum up all dragonflies. It is derived from *libella*, the Latin term for a 'T'-shaped carpenter's level. French scientist Guillaume Rondelet (1505–1566) was the first to use this term in his 1554 book *Libri de piscibus marinis*, applying it to a Hammerhead Shark (*Libella marina*) due to its 'T'-shaped head, and later to a damselfly larva (*Libella fluviatilis*) for presumably similar reasons. Other scientists, including Linnaeus, then applied the name *Libellula* to all odonates.

Robin Ngiam

Libellula quadrimaculata Linnaeus, 1758, the Four-spotted Chaser, a common skimmer found in temperate Europe, Asia and North America.

Trumpet Tail *Acisoma panorpoides* Rambur, 1842

Size HWL: 20–22mm; TBL: 27–29mm

Description Small dragonfly that is easily recognized by its uniquely shaped abdomen, with segments 2–5 markedly swollen, followed by tapering tip. Male has azure-blue eyes. Thorax and abdomen black with numerous light blue lines and markings. Abdominal segments 8–10 black. Appendages white. Female similarly shaped, but with pale green eyes and light green markings.

Habitat & Habits Found at grassy edges of drains, ponds, lakes and reservoirs; also at grassy banks of streams flowing through forest clearings. Perches very low among grasses and other marginal vegetation, its size and colours – especially the female – rendering it fairly inconspicuous. Fierce predator, consuming other dragonflies of near equal size, such as the Black-tipped Percher (p. 241).

Presence in Singapore Recorded in the nature reserves, nature parks, Singapore Botanic Gardens, Pulau Ubin, and urban parks such as Toa Payoh Town Park, Jurong Lake Gardens and Bishan-Ang Mo Kio Park.

Etymology Generic epithet may be a derivation of *akís* (Greek for 'needle'), possibly referring to tapering abdomen. Specific epithet, meaning 'resembling panorpa', refers to abdomen, which resembles that of the European Scorpionfly (*Panorpa communis*).

Distribution Much of tropical and subtropical Asia. Formerly thought to range into Africa, but African look-alikes were recently revised and the genus now contains six species, including *A. attenboroughi*, a Madagascar endemic named in honour of the British naturalist and broadcaster Sir David Attenborough.

National Conservation Status Least Concern; Widespread and Common.

IUCN Red List Status Least Concern.

Larva Fairly typical of family, though smaller (c. 10mm) than most, with trapezoid-shaped head, stout, robust body and moderately long legs.

Marcus Ng

Dorsal view of a male, showing the characteristic shape of the abdomen.

Marcus Ng

Male, perched low amid pondside vegetation.

Marcus Ng

Female, with its mostly green hues.

Marcus Ng

*A scorpionfly (*Neopanorpa *species) photographed in Vietnam, showing the typical abdomen shape of the family.*

Tang Hung Bun

Pair in wheel.

BLUE ADJUTANT *Aethriamanta aethra* Ris, 1912

Size HWL: 23–25mm; TBL: 27–30mm

Description Small, dark blue dragonfly with compact body and largish head. Mature male has dark brown and blackish eyes. Thorax covered by royal blue pruinescence, with irregular dark markings. Abdomen mostly blue, except for segments 7–10, which are black. Hindwing has fairly large, dark brown basal patch. Female similar, but eyes paler and abdominal segments 6–10 dark. Young adults of both sexes have pale yellow and black markings, which become obscured by blue pruinescence as they mature. Distinguished from the Blue Dasher (p. 223) by smaller size, darker colours (Blue Dasher has brown on thorax) and more open wing venation. Similarly sized Mangrove Dwarf (p. 290) is a lighter shade of blue, has totally hyaline wings and seldoms occurs far from the canopy of mangrove forests.

Habitat & Habits Found around marshes, open swamp forests and well-vegetated ponds with ample floating plants in open country; also landward fringes of mangroves. Widespread but elusive and fugitive insect. Usually evident only during the hottest hours of the day, when males take up positions by the water, often in an obelisk position, and furiously chase other dragonflies that come close.

Presence in Singapore Recorded in several locations, including the Sungei Buloh Wetland Reserve, Lim Chu Kang, Kranji Marshes, Nee Soon Swamp Forest, Dairy Farm Nature Park (Singapore Quarry), Bishan-Ang Mo Kio Park, Pasir Ris and the Istana.

Etymology Possibly a combination of *aethra* (Latin for 'bright sky') and *amans* ('loving'). Generic epithet may mean 'loving the bright sky', which certainly describes the dragonfly's behaviour, although Kirby, who coined the genus, probably had the insect's colour in mind. Aethra is also the mother of Helios (the sun god) in Greek mythology.

Distribution Parts of Sundaland and Indochina.

National Conservation Status Least Concern; Widespread but Uncommon.

IUCN Red List Status Least Concern.

Larva Typical of genus (see next two species). Similar to larva of the closely related Coastal Glider (p. 255) and found among floating plants such as the Water Lettuce (*Pistia stratiotes*). The genera *Aethriamanta*, *Macrodiplax* and *Urothemis* are often placed in a subfamily called Urothemistinae (baskers), which have very open wing venation and relatively few antenodal crossveins.

Marcus Ng

Mature male in obelisk, showing its strongly tinted hindwing-bases, open wing venation and deep blue hues.

Marcus Ng

Mature male. The extensive tint at the wing-base easily distinguishes this species from the Pond Adjutant (p. 217).

QM Zhou

Young male that has not yet developed its adult colours.

Tang Hung Bun

Mature female hovering over a pond.

Tang Hung Bun

Pair in wheel.

Scarlet Adjutant *Aethriamanta brevipennis* (Rambur, 1842)

Size HWL: 23–25mm; TBL: 25–28mm

Description Small, bright red dragonfly with short, stout abdomen. Easily distinguished from other red dragonflies by small size, compact build and very open wing venation. Male has dark brown and black eyes. Thorax black and hairy. Abdomen relatively short and bright red. Hindleg has red spot on distal end of femur. Wings hyaline, with dark brown, yellow-edged patch at hindwing-bases. Younger males have brownish thorax and orange-yellow abdomen. Female has brown and light green eyes, brownish-yellow body with black markings, and yellow spot on hindleg femur; coloured patch on wing-base more extensive than in males.

Habitat & Habits Found around weedy ponds and marshy edges of reservoirs near forests. Conspicuous but very alert, and can be hard to approach. Most active towards midday, when males perch on emergent vegetation and chase off all other dragonflies, even far larger species, that come close. Females may be found in nearby vegetation or forest edges.

Presence in Singapore First recorded in Singapore in 2004 at the Botanic Gardens, but has since become more widespread and abundant. Recorded in the Central Catchment Nature Reserve, Sungei Buloh Wetland Reserve, the Singapore Botanic Gardens, Pasir Ris Park, Toa Payoh Town Park, Bishan-Ang Mo Kio Park, Kent Ridge Park, Jurong Eco-Park and Pulau Ubin.

Etymology In Latin, *brevipennis* means 'short winged'.

Distribution Mainland Southeast Asia, Sumatra, South Asia and southern China.

National Conservation Status Least Concern; Widespread but Uncommon.

IUCN Red List Status Least Concern.

Larva Small and squat, with relatively large head.

Marcus Ng

Mature male, showing the shortish abdomen and tint at the hindwing-bases.

Marcus Ng

A younger male with a darkening thorax. Note the bright red 'knees' on the hindlegs.

Marcus Ng

Male with an orange abdomen, probably a young individual.

Marcus Ng

The wing-base colours and coloured 'knees' distinguish the female from other female adjutants.

Tang Hung Bun

Pair in tandem.

Marcus Ng

Head-on view of a female, showing the extensive colours of the wing-bases.

Pond Adjutant *Aethriamanta gracilis* (Brauer, 1878)

Size HWL: 22–23mm; TBL: 26–28mm

Description Small, compactly built, light blue dragonfly. Male has dark brown and blackish eyes. Thorax and abdomen covered by powder-blue pruinescence, except for segments 8–10, which are dark. Wings hyaline, with slight yellow-brown tint at hindwing-base. Distinguished from the Blue Adjutant (p. 213) by its paler shade of blue and reduced colour at hindwing-base. Female has dark brown and greenish eyes. Thorax and abdomen rich brown with thick blackish markings. Male often confused with the Blue Dasher (p. 223), but is markedly smaller with more lightly built thorax (without brown markings on sides). Wings also feature more open venation (six antenodal crossveins in forewing versus seven in the Blue Dasher) and are more pointed at tips. All adjutants have very open wing venation, with relatively fewer intercalated veins at rear margins, distinguishing them from similar looking blue skimmers such as the Blue Dasher and Mangrove Dwarf (p. 290), which have denser wing venation.

Habitat & Habits Found at weedy ponds and marshy edges of reservoirs and lakes. Prefers fairly open areas. Absent from closed forests. Males command perches on waterside vegetation around midday and actively chase off other approaching dragonflies. Usually easier to approach than other members of genus. Like the Scarlet Adjutant (p. 215), this species appears to be expanding in distribution.

Presence in Singapore Most widespread and common local adjutant. Recorded in several locations including the Central Catchment and Labrador Nature Reserves, Bukit Batok and Windsor Nature Parks, Singapore Botanic Gardens, Kent Ridge Park, Toa Payoh Town Park, Pasir Ris Park, Bishan-Ang Mo Kio Park and Pulau Ubin.

Etymology In Latin, *gracilis* means 'slender' or 'gracile'.

Distribution Southeast Asia.

National Conservation Status Least Concern; Widespread and Common.

IUCN Red List Status Least Concern.

Larva Typical of genus, small with relatively large head.

Marcus Ng

Lateral view of a male. Note the thorax, which lacks brown, and the black on the abdomen, which is less extensive than in the Blue Dasher.

Marcus Ng

Dorsal view of a male, showing the open wing venation, with just six antenodal crossveins.

Robin Ngiam

Female, showing the extensive dark markings on the abdomen.

Tang Hung Bun

Lateral view of the female.

Tang Hung Bun

Pair in wheel.

GRENADIER *Agrionoptera insignis* (Rambur, 1842)

Size HWL: 28–30mm; TBL: 37–41mm

Description Small-medium dragonfly with thin red abdomen. Eyes dark brown and light green. Wings hyaline and narrow, with rudimentary (not sock-shaped) anal loop. Male synthorax and abdominal segments 1–2 dark (appears metallic green in good light) with irregular mottled yellow markings. Abdomen thin but swollen basally, slightly horizontally compressed and often slightly arched in profile. Segments 3–7 red with black banding; segments 8–10 and appendages black. Older males develop greyish pruinescence that obscures thoracic markings. Female similar to male but abdomen thicker and more swollen at base, with small ventral flaps on segment 8. Abdominal segments 3–7 marked with orange to red; these colours are obscured by blue-grey pruinescence with age. Distinguished from the Handsome Grenadier (p. 221) by smaller size and amount of red on abdomen. Irregular thoracic markings also distinguish it from the Striped Grenadier and Scarlet Grenadier (pp. 259 and 251), which have more well-defined thoracic stripes.

Habitat & Habits Found around slow streams and swampy pools in shaded forests or forest edges; also landward margins of mangroves and coastal swamps. May occur in wooded habitats near urban areas. Widespread species in both mature and disturbed forests, as well as well-vegetated secondary and suburban habitats. Both sexes often encountered on low vegetation along shaded forest trails. Males territorial, occupying prominent perch at their breeding sites and pursuing similarly coloured dragonflies that come close. Conspecific males are met with a mid-air challenge, with both rivals facing and circling each other until one breaks off and leaves.

Presence in Singapore Recorded in the Bukit Timah, Labrador and Central Catchment Nature Reserves, Sungei Buloh Wetland Reserve, Singapore Botanic Gardens and Pulau Ubin. Also occurs in most nature parks as well as urban parks with shaded ponds close to forests, such as Kent Ridge Park, Pasir Ris Park and Yishun Park.

Etymology Generic epithet combines the Latin term for 'damselfly' (*Agrion*) and Greek for 'wing' (*pteron*). Brauer, who coined the genus in 1864, was struck by the narrow hindwings, which resemble the forewings in size and shape, as in damselflies. Specific epithet means 'remarkable' in Latin and refers to the same observation, which is rare among European true dragonflies but not uncommon in many tropical forest-dwelling skimmers.

Marcus Ng

Young male, showing the mottled thoracic markings.

Distribution Tropical Asia and Australasia, with several regional subspecies in need of taxonomic review.

National Conservation Status Least Concern; Widespread and Common.

IUCN Red List Status Least Concern.

Larva Dark brownish with banded legs. Inhabits bottom leaf litter and detritus.

Marcus Ng

Mature male, with pruinescence obscuring his thoracic markings.

Marcus Ng

Young female, lacking the pruinescence of older individuals.

Marcus Ng

Older female with a pruinescent thorax.

Marcus Ng

Very old female with faded colours.

Marcus Ng

Pair in wheel. No courtship is evident and copulation lasts just a minute or so.

Robin Ngiam

The dark brown larva.

HANDSOME GRENADIER *Agrionoptera sexlineata* Selys, 1879

Size HWL: 34–36mm; TBL: 40–44mm

Description Medium-sized dragonfly with unique red and blue markings and long wings that give the impression of a much larger insect. Eyes dark brown and greenish-yellow. Male has deep metallic green thorax with yellow lateral stripes that are obscured by blue-grey pruinescence in older individuals. Abdominal segments 2–3 powder-blue above, set against bright red of segments 6–7; red on segment 5 usually faint. Wings narrow (hindwing lacks sock-shaped anal loop) and hyaline, with slightly darkened tips. Female has similar thoracic patterns, but markedly larger and more robust, with ventral flaps beneath abdominal segment 8. Abdomen lacks blue at base but has orange-yellow dorsal and lateral markings, and orange to red 'tail-light' on segments 6–7. May be mistaken for female Dark-tipped Forest Skimmer (p. 237), which has a similar profile, from afar.

Habitat & Habits Found at swampy pools in closed forests. Much less common than the Grenadier (p. 219), preferring more mature forests, and rather local. Males guard small, sometimes ephemeral pools filled with leaf litter and plant debris in swampy forests, perching on an overhanging branch from midday until mid-afternoon, when the sun reaches the forest floor. Both sexes may also bask or forage along semi-shaded forest trails.

Presence in Singapore Recorded in various forested locations, including the Bukit Timah and Central Catchment Nature Reserves, Windsor, Thomson and Chestnut Nature Parks, the Botanic Gardens (formerly) and Pulau Ubin. Described from a specimen collected by Wallace in Singapore.

Etymology Specific epithet, which means 'six-lined', probably refers to three stripes on each side of the thorax.

Distribution Singapore, Peninsular Malaysia, Borneo and Sumatra.

National Conservation Status Least Concern; Widespread but Uncommon.

IUCN Red List Status Least Concern.

Larva Very dark with banded legs. Found in bottom leaf litter.

Marcus Ng

Dorsolateral view of the male, showing the red-and-blue livery and narrow wings (the anal loop is not sock shaped, unlike the Scarlet Grenadier's).

Marcus Ng

Lateral view of the male, showing the thoracic markings.

Marcus Ng

Typical female, with red on abdominal segments 6–7.

Marcus Ng

Female, possibly an older individual, with pale colours on abdominal segment 7.

Marcus Ng

Dorsal view of a male perched by a forest trail.

Robin Ngiam

The very dark larva.

BLUE DASHER *Brachydiplax chalybea* Brauer, 1868

Size HWL: 24–27mm; TBL: 33–35mm

Description Small-medium blue dragonfly with boxy thorax. Eyes brown and light green. Male's thorax covered by powder-blue pruinescence, interspersed with light brown lateral markings. Abdomen powdery-blue, turning black from segment 7. Abdomen-tip and appendages black. Wings hyaline, with yellow-brown tint at hindwing-base. Female brownish-yellow, with dark dorsal markings on abdomen and dark abdominal tip. Hindwing-base clear. Older females may develop bluish pruinescence and resemble males. Distinguished from other smallish blue skimmers such as the Black-tailed Dasher, Pond Adjutant and Mangrove Dwarf (pp. 225, 217 and 290) by larger size and more robust build, in particular the boxy thorax bearing brown side markings. Also differs from blue-coloured adjutants by denser wing venation (see Pond Adjutant).

Habitat & Habits Found at well-vegetated ponds and drains in urban areas and disturbed forests; also marshy edges of forests and reservoirs, and landward sides of mangroves. Favours exposed areas and clearings; seldom found in closed forests. Males highly territorial, actively confronting rivals from midday to mid-afternoon. They are less active later in the day, but remain at their territories when most other diurnal dragonflies have left the scene. Females much less commonly seen, coming down to water only to breed, but may be encountered in vegetation further away from breeding sites.

Presence in Singapore Widespread across the island in urban, suburban and forested habitats.

Etymology Generic epithet combines the Greek for 'short' (*brachy*), referring to thickish abdomen, and *diplax*, an element applied by Brauer to some genera of skimmers, such as *Macrodiplax* and *Pachydiplax*. *Diplax*, which means 'bi-lobed' (for two lobes on prothorax) is an invalid synonym of *Sympetrum*, a widespread genus of skimmers. Specific epithet is Latin for 'having the colour of tempered steel'.

Distribution Widespread in tropical Asia.

National Conservation Status Least Concern; Widespread and Common.

IUCN Red List Status Least Concern.

Larva Light to darkish brown with long, hairy legs.

Marcus Ng

Lateral view of the male, showing the boxy thorax with brownish streaks, which distinguish this species from the smaller Pond Adjutant.

Marcus Ng

Dorsal view of a male. Note the denser wing venation (with seven antenodal crossveins) compared to the Pond Adjutant.

Marcus Ng

Typical female.

Marcus Ng

Andromorph female. Note the vulvar scale near the abdomen-tip.

Marcus Ng

Close-up of the male's head and thorax.

Robin Ngiam

The brown larva.

BLACK-TAILED DASHER *Brachydiplax farinosa*
Krüger, 1902

Size HWL: 21–25mm; TBL: 28–31mm

Description Smallish, powder-blue dragonfly that is fairly common elsewhere in the region but very rare locally. Male similar to the Blue Dasher (p. 223), but slightly smaller, with less boxy thorax that lacks extensive brown of the other species. Eyes also darker. Abdomen narrower, with more extensive black that starts from segment 6. Wings hyaline, but some individuals may have slight basal tint. Forewing has 8–9 antenodal crossveins, versus seven in Blue. May be confused with the Pond Adjutant (p. 217), but black on abdomen more extensive. Also, wings usually lack basal tint and are more densely veined; the Pond Adjutant has just six antenodal crossveins in forewing. Female has light brown and green eyes. Thorax has well-defined black and pale yellow stripes, turning blue-grey with age. Abdomen orange-yellow to reddish with black markings, becoming darker with age.

Habitat & Habits Found at ponds and marshes near forests; also in swampy forests. Prefers less exposed habitats than the Blue Dasher and Pond Adjutant. Smallish blue dragonflies in swampy, forested habitats should be examined carefully to determine if they are this species.

Presence in Singapore Very rare dragonfly in Singapore, despite being fairly common in Peninsular Malaysia and Thailand. Recorded in Murnane Reservoir in 1971 and Sime Road in 1989.

Etymology Specific epithet means 'floury' in Latin, a reference perhaps to male's powder-blue colour.

Distribution Sundaland and mainland tropical Asia.

National Conservation Status Endangered; Restricted and Very Rare.

IUCN Red List Status Least Concern.

Larva Appearance typical of genus. Very similar to that of Blue Dasher (final instar 17mm or more), but smaller in size (final instar 16mm).

Marcus Ng

Male in Gopeng, Malaysia. Note the less boxy thorax and greater amount of black on the abdomen compared to the Blue Dasher.

Marcus Ng

Male photographed in Kaeng Krachan, Thailand. Note the forewing's eight antenodal crossveins, which distinguish this species from the Blue Dasher (seven crossveins) and Pond Adjutant (six crossveins).

Marcus Ng

Female photographed in Gopeng, Malaysia.

PIXIE *Brachygonia oculata* (Brauer, 1878)

Size HWL: 17–19mm; TBL: 21–24mm

Description Small dragonfly with unmistakable, highly contrasting patterns. Wings hyaline and rounded at tips. Male's eyes dark brown and blackish, with paler edging. Thorax has pale brown and yellowish markings. Abdomen orange, with powdery bluish-white on dorsum of segments 3–5. Segments 8–10 and appendages black. Female has paler eyes and colours; abdomen brownish and lacking male's distinctive bluish-white markings.

Habitat & Habits Found around shaded or semi-shaded pools and very sluggish streams in swampy forests or forest edges. Males guard territories at small pools filled with leaf litter and other detritus, perching for long periods. Both sexes may also feed by forest trails near swampy areas.

Presence in Singapore Wallace collected two males in Singapore in 1854. Currently known from two populations from the Western Catchment. Vagrant male was recorded in Admiralty Park in 2015. Species is part of the National Parks Board's Species Recovery Programme to boost the population of threatened species, in this case by introducing the dragonfly to suitable new locations.

Etymology Generic epithet combines the Greek for 'short' (*brachy*) and 'generation' or 'begetting' (*gonia*), probably referring to male's inconspicuous secondary genitalia. Specific epithet (Latin for 'furnished with eyes') refers to prominent compound eyes.

Distribution Sundaland and southern Indochina.

National Conservation Status Endangered; Restricted and Uncommon.

IUCN Red List Status Least Concern.

Larva Very small (final instar 9–10mm); dark brown with long, dark-banded legs. Abdomen oviform. Inhabits plant debris at bottoms of forest pools.

Marcus Ng

Lateral view of the male, showing the conspicuous pale hues of abdominal segments 3–5.

Robin Ngiam

Dorsal view of the male, showing the rounded wing-tips.

Cheong Loong Fah

Lateral view of the female.

Bo Nielsen

Lateral view of female photographed in Pangkor, Malaysia.

Robin Ngiam

Dorsal view of the female.

Max Khoo

The larva showing its banded legs.

Common Amberwing *Brachythemis contaminata* (Fabricius, 1793)

Size HWL: 21–23mm; TBL: 29–31mm

Description Smallish dragonfly with richly coloured wings and rather short, thickish abdomen. Eyes dark brown and yellow-brown (lower portion green in younger individuals), with faint barring on top. Male's thorax brown with obscure markings. Abdomen brownish-orange, with pale, broken dorsal line, which becomes obscured with age. Both wings have basal amber tint that becomes richer after nodus but ends before pterostigma. Pterostigmata orange-red. Female and immature male similar but paler, with hyaline wings marked by amber primary veins and paler pterostigmata. May be confused with young male and female Common Parasols (p. 263), but has a stouter profile and different abdominal markings; note pale broken dorsal line, absent in Common Parasol.

Habitat & Habits Occurs around drains, ponds, lakes, reservoirs and other still or slow-flowing waterbodies. Widespread and tolerant species that inhabits many open and disturbed habitats, including polluted waters. Both sexes can be found perched by the water or on nearby grass verges and bushes. Females and immature males may fly very low at ground level, behaviour that gave rise to common name 'groundlings' for African members of the genus. Active until around sunset.

Presence in Singapore Recorded island-wide at most suitable habitats.

Etymology Generic epithet combines the Greek for 'short' (*brachys*), referring to the stout abdomen, and *themis*, a common element in true dragonfly genera, such as *Neurothemis* and *Rhyothemis*. Themis is the Greek goddess of order and an apt deity for taxonomists. Specific epithet means 'contaminated' or 'stained', perhaps referring to colouration of wings.

Distribution Tropical and subtropical Asia.

National Conservation Status Least Concern; Widespread and Common.

IUCN Red List Status Least Concern.

Larva Wide head relative to thorax; abdomen oviform.

Marcus Ng

Dorsal view of the male, showing the orange wing colours and dorsal markings of the abdomen.

Marcus Ng

Lateral view of the male, showing the shortish abdomen.

Marcus Ng

Lateral view of the female. The abdomen shape and markings distinguish this species from the female Common Parasol.

Marcus Ng

Dorsal view of an older female with duller colours.

SULTAN *Camacinia gigantea* (Brauer, 1867)

Size HWL: 44–48mm; TBL: 53–56mm

Description Largest member of the family Libellulidae. Large and conspicuous dragonfly that suggests a giant Common Parasol (p. 263) but with very different flight and habits. Male has rich brown eyes and a deep red body. Red extends to basal two-thirds of both wings, ending between nodus and pterostigma, and becoming darker at outer edges. Immature male paler, turning from yellow to red with age. Female slightly larger, more yellowish-brown, and with less extensive colour on wings. Andromorphs not uncommon.

Habitat & Habits Occurs around open ponds, quarry lakes and forest edges; also landward sides of mangroves. Strong and fast flyer that probably stays in the canopy when not active at breeding sites. Perches in hanging position, unlike superficially similar Common Parasol and most other smaller skimmers, which perch with the abdomen fairly level. Males actively patrol breeding sites on sunny days, gliding and swooping over smaller skimmers and challenging conspecifics. They seek out nearby perches during overcast periods. Newly arrived females are seized without any apparent courtship displays. Copulation very brief – a minute or so – and almost immediately after mating female oviposits by repeatedly dipping her abdomen into the water to release batches of eggs.

Presence in Singapore Recorded in various forested green spaces, including the Bukit Timah and Central Catchment Nature Reserves and adjacent nature parks, Sungei Buloh Wetland Reserve, the Singapore Botanic Gardens, Coney Island and Pulau Ubin. Also common at the Singapore Zoo, which has many suitable waterbodies.

Etymology Generic epithet may be derived from *kamakias*, a Greek term for 'made of reed or cane', referring to the parallel-sided abdomen. Kirby, who coined the genus in 1889, described the abdomen as being 'stout, not thickened at the base, the sides nearly parallel'. Specific epithet means 'gigantic' in Latin.

Distribution Tropical Asia from northern India to New Guinea.

National Conservation Status Least Concern; Widespread but Uncommon.

IUCN Red List Status Least Concern.

Larva Similar to larva of the Water Monarch (p. 245), but more elongated and lacking abdominal spines. Thought to be semi-pelagic in behaviour.

Marcus Ng

Mature male, showing the extensive wing colours and dense venation.

James Holden

Male Sultan (right), showing its size compared to a male Common Parasol (left). Photo taken in Vietnam.

Marcus Ng

Young male photographed in Gopeng, Malaysia.

Yeo Suay Hwee

Typical female.

Phil Benstead

Andromorph female. Note the short cerci on the abdomen.

Larva.

Green-eyed Percher *Chalybeothemis fluviatilis* Lieftinck, 1933

Size HWL: 21–22mm; TBL: 29–30mm

Description Smallish, lightly built, dark blue dragonfly with green eyes and fairly long legs. Male's thorax dark steely-blue with obscure blackish markings. Abdomen blackish and thin, with slightly swollen base. Female similar, but with brown on dorsum of thorax and brownish-yellow tint at wing-bases.

Habitat & Habits Found around weedy banks of streams, open ponds and lakes, and marshy edges of reservoirs close to forests. Rather inconspicuous dragonfly of open habitats close to forests. Perches on low grasses and reeds by the water's edge. Hard to approach, flying off and skimming low over banks when disturbed.

Presence in Singapore Recorded at MacRitchie Reservoir, Windsor Nature Park, HortPark, the Western Catchment and Pulau Ubin. Local and unpredictable in occurrence.

Etymology Generic epithet combines *chalybeus* ('having the colour of tempered steel') with *themis*, a common suffix in libellulid genera. Specific epithet means 'of a river' in Latin.

Distribution Parts of Sundaland, southern Thailand and Cambodia.

National Conservation Status Vulnerable; Restricted and Uncommon.

IUCN Red List Status Least Concern.

Larva Unknown for genus.

Marcus Ng

Dorsal view of the male. Note the green eyes and hyaline wings.

Robin Ngiam

Male, showing the thin abdomen and long legs.

Tang Hung Bun

Lateral view of a male.

Tang Hung Bun

Pair in wheel.

LINED FOREST SKIMMER *Cratilla lineata*
(Brauer, 1878)

Size HWL: 36–38mm; TBL: 46–48mm

Description Moderately large, metallic blue dragonfly restricted to forests. Eyes brown and yellowish-green. Male's thorax dark blue with thin yellow lateral and dorsal stripes, which become obscured by greyish bloom in older individuals. Abdomen bluish-black, with thin broken yellow line running along dorsum until segment 8. Yellow becomes obscured with age, as wing-bases and basal segments of abdomen develop a light grey-blue pruinescence. Wings largely hyaline but may be slightly darkened at tips in some individuals (though never as strongly as in the Dark-tipped Forest Skimmer, p. 237). Female similarly coloured but more robust. Abdomen has small ventral flaps on segment 8. Wing-tips usually darkened (more so than in male). Distinguished from female Dark-tipped by slightly smaller size, less robust build and thin yellow stripes on thorax.

Habitat & Habits Found around small, shallow pools in closed forests or swampy forests. Forest-dependent dragonfly that is most abundant near northern coast and nearby islands. Males guard small, shallow pools in shaded forests. Both sexes may be encountered on masses of fallen branches in clearings or along trails.

Presence in Singapore Recorded in various locations since its first local discovery in 2006, including the Central Catchment Nature Reserve, Sungei Buloh Wetland Reserve, Punggol, Pasir Ris, Sembawang, Coney Island, the Night Safari, Pulau Ubin and Pulau Tekong.

Etymology Generic epithet comes from *cratis*, Latin for 'lattice', and may refer to dense wing reticulation (a feature not unique to the genus, however). Specific epithet, Latin for 'lined', refers to abdominal markings.

Distribution South and Southeast Asia, and southern China. Three subspecies recognized within its range.

National Conservation Status Least Concern; Widespread but Uncommon.

IUCN Red List Status Least Concern.

Larva General appearance similar to larva of Dark-tipped Forest Skimmer.

Marcus Ng

Lateral view of a typical male, showing the yellow thoracic stripes.

Marcus Ng

Lateral view of the female, showing its thoracic markings and ventral flaps, which are smaller than those of the female Dark-tipped Forest Skimmer.

Robin Ngiam

Male with slightly darkened wing-tips and pruinescence at the abdomen-base.

Marcus Ng

Dorsal view of a young female. Note the yellow dorsal stripe along the abdomen, which is absent in the Dark-tipped Forest Skimmer.

Marcus Ng

Older female with pruinescence covering most of body.

Marcus Ng

Lateral view of an older female.

Dark-tipped Forest Skimmer *Cratilla metallica* (Brauer, 1878)

Size HWL: 36–38mm; TBL: 46–51mm

Description Fairly large dragonfly with unmistakable colours. Eyes dark brown and greenish-yellow. Wing-tips of both sexes dark. Male's thorax and abdomen metallic dark blue except for abdominal segments 3–4, which are lighter slate-blue. Dorsum of thorax has broken yellow line that does not continue into abdomen as in the Lined Forest Skimmer (p. 235). Female similar but more robust, and abdomen lacks lighter blue on segments 3–4. Younger individuals may have yellow lines on sides of thorax.

Habitat & Habits Found in closed forests. Fairly widespread and common in both mature and disturbed forests, where both sexes can be encountered at shaded trails, forest edges, and masses of fallen branches in clearings. Males guard small, shallow forest pools or phytotelmata for up to several days. Females that approach the breeding site are quickly seized, if willing, and after a few minutes of copulation, begin to lay eggs by rapidly flipping the abdomen at the surface to fling wetted eggs at the water's edge.

Presence in Singapore Recorded in many forested locations, including the Bukit Timah and Central Catchment Nature Reserves and adjacent nature parks such as Windsor, Dairy Farm and Thomson Nature Parks; also in the Singapore Botanic Gardens (formerly), Jurong Eco-Garden, Bukit Brown and Pulau Ubin.

Etymology Specific epithet describes the highly metallic body, an uncommon feature in skimmers.

Distribution Sundaland, Thailand, the Philippines, Nepal and India (possibly).

National Conservation Status Least Concern; Widespread and Common.

IUCN Red List Status Least Concern.

Larva Dark, hairy and rather elongated for a skimmer, living amid bottom debris and under leaf litter in forest pools and phytotelmata. Distinctive broad dark stripes on front of labium. Aggressive predator, taking tadpoles that share its forest pools.

Marcus Ng

Dorsal view of a male, showing the lighter blue markings on abdominal segments 3–4.

Marcus Ng

Lateral view of a male, showing the distinctive eye and facial colours.

Marcus Ng

The female's abdomen, especially the terminal segments, is more robust than that of the Lined Forest Skimmer.

Marcus Ng

Lateral view of a female. Note the pronounced ventral flaps.

Marcus Ng

Pair in wheel. Copulation lasts for just a couple of minutes.

Marcus Ng

The dark larva, sharing a large concrete tank with frog tadpoles.

Robin Ngiam

Close-up of the larva, showing its dark frontal stripes.

COMMON SCARLET *Crocothemis servilia* (Drury, 1773)

Size HWL: 31–33mm; TBL: 40–43mm

Description Medium-sized red dragonfly with conspicuous habits. Male has red eyes (paler below) with white edging. Thorax and abdomen bright red, with thin black dorsal stripe running along length of abdomen. Appendages red. Wings hyaline except for brownish-yellow patch at hindwing-base. Female and young males have a brownish-yellow body (also with black dorsal stripe), and light brown and pale green eyes. May be confused with similarly sized Common Redbolt (p. 292), which lacks black dorsal stripe and has eyes that touch at just a single point (best seen from above) rather than broadly. May also be confused with the Scarlet Skimmer (p. 280), which is larger, has brown-grey eyes and also lacks black abdominal stripe. Both *Crocothemis* and *Rhodothemis* species have an incomplete distal antenodal crossvein in the forewing, separating these genera from *Orthetrum* species, which have a complete distal antenodal crossvein.

Habitat & Habits Found at ponds, marshes, reservoirs and drains in open and disturbed habitats; also forest edges and well-vegetated parks and gardens. May occur in high densities, especially at freshly established waterbodies. Females and young males may wander far from water to forage at forest edges and in urban parks, gardens and fields. Mature males conspicuous at breeding sites, occupying a prominent perch and actively pursuing other red dragonflies.

Presence in Singapore Recorded island-wide. Probably the most widespread and abundant medium-sized red skimmer locally.

Etymology Generic epithet combines the Greek for 'saffron' (*krokos*) with *themis*, a common appellation for libellulid dragonflies, referring to orange-yellow tint in hindwing-bases of members of this genus. Servilia was the mistress of the Roman dictator Julius Caesar and mother of Brutus, Caesar's assassin.

Distribution Tropical and subtropical Asia from the Middle East to New Guinea.

Marcus Ng

Dorsal view of a roosting male, showing incomplete antenodal crossvein in the forewing.

Established in Florida, Hawaii and Jamaica, probably through larvae transported with aquarium plants.

National Conservation Status Least Concern; Widespread and Common.

IUCN Red List Status Least Concern.

Larva Squat and dark. Can thrive in stagnant waterbodies such as household water-storage jars. Has been tested as an effective means for biological control of mosquito larvae in Myanmar.

Marcus Ng

Lateral view of a female. Note the short cerci on the abdomen.

Marcus Ng

Young male.

Robin Ngiam

The dark and squat larva.

Black-tipped Percher *Diplacodes nebulosa* (Fabricius, 1793)

Size HWL: 18–19mm; TBL: 23–25mm

Description Small dragonfly distinguished by its dark wing-tips. The only other local skimmer with such dark wing-tips is the much larger, forest-dwelling Dark-tipped Forest Skimmer (p. 237). Male has dark brown and greenish eyes with dark flecks. Thorax and abdomen dark blue. Appendages pale. Wings hyaline except for black at tips, starting from pterostigma. Female has light brown and green eyes. Thorax and abdomen greenish-yellow, with dark streaks and strong dark dorsal stripe. Wing-tips clear. Young male similar to female, with dark wing markings showing up with age. Females and young males may be confused with females and young males of the Blue Percher (p. 243), which is larger and lacks broad black dorsal stripe on thorax.

Habitat & Habits Found around marshes and shallow edges of ponds in open country. Locally abundant species in suitable habitat, perching low on grasses and reeds by the water or marshy banks. Males highly territorial – their wings clash audibly during bouts with rivals.

Presence in Singapore Recorded in various locations such as MacRitchie Reservoir, East Coast Park, Labrador, Tuas, Kranji, Pasir Ris, Kent Ridge Park, and the Singapore Botanic Gardens.

Etymology *Diplacodes* means 'resembling Diplax'. Kirby coined this genus for dragonflies formerly placed in the genus *Diplacina* and the invalidated genus *Diplax* (see also etymology of *Brachydiplax*). Specific epithet, from *nebulosus* (Latin for 'misty' or 'foggy'), may refer to the dark wing-tips.

Distribution Tropical Asia and Australasia, from India to northern Australia.

National Conservation Status Least Concern; Widespread but Uncommon.

IUCN Red List Status Least Concern.

Larva Typical libellulid in appearance.

Marcus Ng

Mature male.

Marcus Ng

Female, showing the dark dorsal stripe, which distinguishes this species from the Blue Percher.

Marcus Ng

Young male with darkening wing-tips.

Marcus Ng

Very young male with hyaline wings.

Tang Hung Bun

Pair in wheel.

BLUE PERCHER *Diplacodes trivialis* (Rambur, 1842)

Size HWL: 22–23mm; TBL: 29–32mm

Description Small dragonfly that is widespread and common but usually inconspicuous as it stays close to the ground. Male has azure-blue eyes. Thorax and abdomen largely chalky-blue except for abdominal segments 8–10, which are black. Appendages white. Female has brown and pale green eyes. Thorax pale green with black markings. Abdomen mostly pale green with dark dorsal line. Distinguished from the female Black-tipped Percher (p. 241) by lack of dark dorsal line on thorax. Immature male similar to female. Females and young males may also be confused with the Variegated Green Skimmer (p. 278), which is much larger and has a more slender abdomen with a swollen base.

Habitat & Habits Found around marshes, ponds and even temporary roadside pools in open and disturbed habitats. Wide-ranging, tolerant species that can be found at unlikely sites far from fresh water, such as urban parks, fields, hilltops, beach vegetation and even rooftop gardens in housing estates. Often the first dragonfly to colonize cleared construction land. Both sexes perch on bare ground or on low vegetation and are inconspicuous, flying low when disturbed.

Presence in Singapore Recorded island-wide.

Etymology Specific epithet means 'common' in Latin, and aptly describes this dragonfly.

Distribution Tropical and subtropical Asia and Australasia from India to Australia and Fiji.

National Conservation Status Least Concern; Widespread and Common.

IUCN Red List Status Least Concern.

Larva Typical libellulid in appearance.

Marcus Ng

Mature male, perched close to the ground.

Marcus Ng

Young male, with greenish hues like the female.

Marcus Ng

Typical female, perched low among grasses in a housing estate.

Marcus Ng

Older female, with blue pruinescence obscuring markings.

Marcus Ng

Pair in wheel by a rural track.

WATER MONARCH *Hydrobasileus croceus* (Brauer, 1867)

Size HWL: 42–45mm; TBL: 46–49mm

Description Large, highly aerial dragonfly with distinctive hindwing markings. Eyes orange-brown, paler below. Male's thorax and abdomen orange-brown, with dark dorsal markings on segments 4–10. Appendages dark. Hindwing-base greatly expanded, with irregular dark brown patch extending from anal angle to 'heel' of sock-shaped anal loop. Female similar but with paler abdomen.

Habitat & Habits Occurs around well-vegetated ponds and lakes in parks and open country; also open areas of forested reservoirs and forest edges. Conspicuous insect that actively patrols ponds, lakes and reservoir edges on sunny days, often gliding and sailing above smaller skimmers. During cloudy weather, may be found perched in hanging position in dense vegetation. Unusually among true dragonflies, this species oviposits while in tandem, a form of contact guarding.

Presence in Singapore Recorded in many urban and semi-urban locations such as the Bukit Timah and Central Catchment Nature Reserves and adjacent nature parks, Sungei Buloh Wetland Reserve, Bukit Batok Nature Park, Tampines Eco-Green and the Singapore Botanic Gardens.

Etymology Generic epithet combines the Greek for 'water' (*hydro*) and 'king' (*basileus*). Specific epithet means 'saffron coloured' (see also etymology of *Crocothemis*).

Distribution Tropical and subtropical Asia.

National Conservation Status Least Concern; Widespread and Common.

IUCN Red List Status Least Concern.

Larva Pale greenish and somewhat translucent, with long, backwards pointing spines on abdomen. Pelagic among aquatic vegetation just under the water's surface.

Marcus Ng

Male, showing the distinctive hindwing markings.

Marcus Ng

Lateral view of a male perched at a forest edge.

Marcel Finlay

Males patrol large waterbodies during sunny weather and are easily recognized by their wing markings.

Marcus Ng

Female clinging to a bush during cloudy weather.

Tang Hung Bun

Pair in tandem.

Robin Ngiam

The pale green larva showing the long abdominal spines.

WHITE-TIPPED DEMON *Indothemis carnatica* (Fabricius, 1798)

Size HWL: c. 29mm; TBL: c. 31mm

Description Small-medium blue dragonfly with fugitive behaviour. Wings hyaline, except for small dark patch at hindwing-base. Eyes of male dark brown and dark greyish-blue. Thorax blue with distinctive bulge on venter (lower rear portion just before abdomen). Abdomen mostly blue with darker dorsal markings that become more extensive towards tip. Segments 9–10 black. Upper appendages black with white proximal end; lower appendage mostly white with black tip. Female has light brown and pale green eyes. Thorax pale yellow with thickish dark stripes on dorsum and thinner stripes on sides. Abdomen light brownish-yellow with dark dorsal and lateral markings that become thicker towards tip. The closely related Restless Demon (p. 249) has a much larger hindwing-patch and males are markedly darker in colour; also, forewing of White-tipped has 8–9 complete antenodal crossveins compared to 10–12 in Restless. Female distinguished from the female Indigo Dropwing (p. 312) by thoracic markings (White-tipped has thinner lateral stripes) and bulging venter. Also distinguished from the female Crimson Dropwing (p. 310) by thoracic markings (less well-defined and incomplete in Crimson Dropwing) and less extensive hindwing tint.

Habitat & Habits Found at weedy ponds, lakes and marshes in open country or near forests. Males active at breeding sites towards noon, flying rapidly over the water in pursuit of rivals, or perching on waterside vegetation.

Presence in Singapore This rather fugitive and inconspicuous species was first recorded locally near Pekan Quarry at Pulau Ubin in 2018, followed by sightings of a single female on the main island (Upper Seletar Reservoir and Marsiling Park) in 2021. Its appearance may mark a southwards dispersal of the species down the Malay Peninsula.

Etymology In Latin, prefix *Indo-* means 'of India', from where this genus was first described. Specific epithet may refer to Carnatic region of India along its southeastern coast (now Tamil Nadu and southern Andhra Pradesh).

Distribution South Asia and mainland Southeast Asia.

National Conservation Status Critically Endangered; Restricted and Very Rare.

IUCN Red List Status Least Concern.

Larva To the best of the authors' knowledge, larva unknown for this genus.

Michael Soh

Lateral view of the male, showing the thoracic markings and white and black anal appendages.

Michael Soh

Lateral view of the male, showing the bulging venter on the underside of the thorax.

Marcus Ng

Lateral view of the female, showing the thoracic markings and bulging venter.

Marcus Ng

Dorsal view of the female, showing the abdominal markings and slight wing-base tint (compare with the female Restless Demon).

Restless Demon *Indothemis limbata* (Selys, 1891)

Size HWL: 26–29mm; TBL: 31–36mm

Description Small-medium, very dark dragonfly with fugitive habits. Male has dark bluish-black eyes. Thorax dark blue, appearing black from a distance. Abdomen and appendages dark blue, with darker dorsal markings. Young males may have yellowish streaks on abdomen. Hindwing-base has small but prominent dark brown patch that ends before rear margin. Wing-tips slightly darkened. Males may be confused with the Indigo Dropwing (p. 312) but that species lacks darkened wing-tips, and hindwing basal patch is more irregular in shape. Habitat preferences also differ, with the Indigo Dropwing preferring streams. Larger hindwing-patch and colour of appendages also distinguishes this species from the White-tipped Demon (p. 247). Female yellowish with dark markings, including broad dark dorsal stripe along abdomen flanked by two thinner stripes. Hindwing-base has yellow-brown tint, more extensive than in the White-tipped Demon.

Habitat & Habits Found around open marshes and reservoir edges with emergent plants (especially sedges). Elusive and fast-flying species that shows up on the brightest of days. Males occupy a sunny perch by the water's edge and actively chase after other dragonflies. Fugitive and unpredictable in occurrence.

Presence in Singapore Recorded from MacRitchie and Upper Peirce Reservoirs; also sporadic sightings at Pasir Ris Park, Labrador Nature Reserve and Sungei Buloh Wetland Reserve. Formerly recorded at Marina East and Marina South. There are records from Normanton Park and the National University of Singapore's Bukit Timah campus in the 1980s.

Etymology Specific epithet means 'edged' or 'bordered' in Greek, and may refer to prominent dark hindwing-patch.

Distribution South Asia and mainland Southeast Asia.

National Conservation Status Vulnerable; Restricted and Uncommon.

IUCN Red List Status Least Concern.

Larva Unknown.

Marcus Ng

Lateral view of a male, showing the dark thorax, slightly darkened wing-tips and extensive dark patch on the hindwing-base.

Marcus Ng

Female at Sembawang Beach, showing its thoracic and abdominal markings (compare with the female White-tipped Demon).

Robin Ngiam

Dorsal view of a male.

Dennis Farrell

Female photographed in Thailand. Note the more extensive hindwing-base tint compared to that of the White-tipped Demon.

Marcus Ng

Older female with darker colours.

Scarlet Grenadier *Lathrecista asiatica* (Fabricius, 1798)

Size HWL: 34–37mm; TBL: 44–47mm

Description Medium-sized, red and blue dragonfly with very straight abdomen. Male has brown and grey-blue eyes (greenish when young). Synthorax dark brown, with pale yellow stripes and streaks on sides, often obscured by blue-grey pruinescence in older individuals. Abdomen thin, straight and mostly bright red except for segments 9–10 and appendages, which are black. Wings hyaline and narrow, but hindwing has sock-shaped anal loop (the only local grenadier to have this). Female similar but abdomen thicker and mostly yellow-brown to pale red. Younger adults of both sexes may have slightly darkened wing-tips. Easily distinguished from the Grenadier (p. 219) by larger size, more well-defined (rather than mottled) thoracic markings and shape of abdomen (straighter and not horizontally compressed). The two species often occur together. Female may be confused with the Striped Grenadier (p. 259), but distinguished by its larger size, sock-shaped anal loop and thoracic markings: Striped has fairly complete yellow stripes, whereas Scarlet has shorter yellow streaks in between its stripes. Older females may be confused with the Common Chaser (p. 286), but the latter has prominent ventral flaps on abdominal segment 8.

Habitat & Habits Occurs at marshes and swamps near coasts and back mangroves, but also inland forests. Breeds in somewhat shaded forest pools. Males occupy breeding sites around midday, when they actively pursue rivals. Both sexes often encountered foraging at sunlit forest edges, and in clearings and trails.

Presence in Singapore Recorded in many locations, including the Bukit Timah, Central Catchment and Labrador Nature Reserves, adjacent nature parks, Sungei Buloh Wetland Reserve, Coney Island, Admiralty Park, Kent Ridge Park, Pasir Ris Park, the Singapore Botanic Gardens and Pulau Ubin.

Etymology Generic epithet combines the Greek for 'hidden' (*lathraíos*) and 'basket' (*cista*), possibly referring to male's inconspicuous secondary genitalia. English name of this species, as well as of *Agrionoptera* and *Nesoxenia* species, stems from their red and blue livery, which

Marcus Ng

Dorsal view of a male, showing the narrow wings, which contain a sock-shaped anal loop.

recalls the uniforms of nineteenth-century grenadiers, soldiers who specialized in throwing grenades at frontlines of battlefields.

Distribution Tropical Asia and Australasia, with several subspecies requiring taxonomic review.

National Conservation Status Least Concern; Widespread and Common.

IUCN Red List Status Least Concern.

Larva Typical libellulid, similar in appearance to that of the Handsome Grenadier (p. 221) but smaller.

Marcus Ng

Mature male, showing the fairly straight abdomen and blue-grey pruinescence on the thorax.

Marcus Ng

Young male with darkened wing-tips, showing the thoracic markings.

Marcus Ng

Typical female with extensive red on the abdomen.

Marcus Ng

Younger female with paler colours and darkened wing-tips.

Marcus Ng

Older female, with pruinescence obscuring the thoracic markings.

BOMBARDIER *Lyriothemis cleis* Brauer, 1868

Size HWL: 33–36mm; TBL: 40–45mm

Description Medium-sized, dull red dragonfly with white face and metallic head markings. Eyes of both sexes dark brown. Male's thorax brown, darker on dorsum, with pale dorsal markings. Wings hyaline. Abdomen and appendages dull crimson, never as bright as those of similarly sized red *Orthetrum* species, which also lack shiny facial colours. Female similar but abdomen thicker and more brownish to orange-brown.

Habitat & Habits Occurs in mature dipterocarp forests. Breeds in phytotelmata such as large bamboo stumps and buttress pans and cavities of large trees. Males guard such breeding sites and mate with visiting females. Both sexes may be seen along forest trails close to areas with mature trees or bamboo groves.

Presence in Singapore Restricted to the Bukit Timah and Central Catchment Nature Reserves, and adjacent nature parks such as Dairy Farm and Windsor Nature Parks. First collected by Wallace in Singapore in 1854.

Etymology Generic epithet refers to male's upper abdominal appendages, which resemble curved arms of a lyre (a 'U'-shaped harp). Cleïs is the mother of Sappho, a Greek poet.

Distribution India, Thailand, Sundaland, the Philippines and Sulawesi.

National Conservation Status Endangered; Restricted and Rare.

IUCN Red List Status Least Concern.

Larva Typical libellulid in appearance. Very dark with spiny legs. Usually the top predator in local phytotelm systems, feeding on aquatic larvae of other insects such as mosquitoes and crane flies. Resistant to desiccation, allowing it to survive short dry spells.

Marcus Ng

Dorsal view of male photographed in Maliau Basin, Sabah, Malaysia. Note the distinctive pale markings on the dorsum of the thorax.

Martin Kennewell

Male photographed in Bukit Timah. Note the dull red abdomen and bright facial colours.

Robin Ngiam

Male showing the bright facial colours.

Marcus Ng

Female on Wallace Trail. Note the metallic face and pale dorsal markings on the synthorax.

Marcus Ng

Lateral view of female on Wallace Trail. Note the very dark eyes.

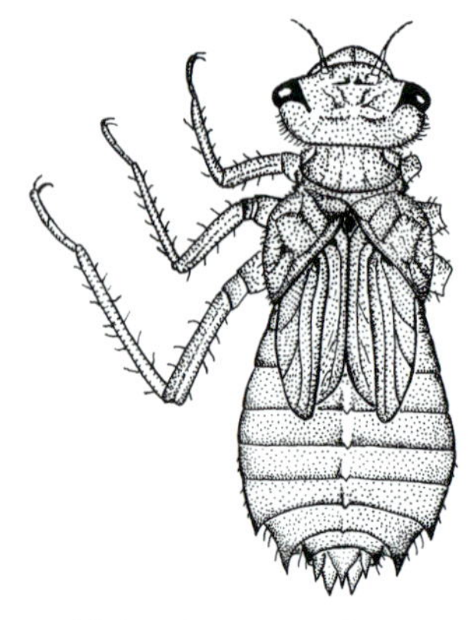

Albert G. Orr

Drawing of the larva, showing its spiny legs.

Coastal Glider *Macrodiplax cora* (Brauer, 1867)

Size HWL: 31–33mm; TBL: 40–43mm

Description Medium-sized red dragonfly with large head and long dark legs. Wings rather pointed, with very open venation, and hindwing-base expanded, giving the impression of a shortish abdomen. Hindwings have very slight yellow-brown tint at bases. Male has brownish-red and dark brown eyes. Thorax dark brown with faint black markings. Abdomen red with dark broken dorsal line. Appendages pale red. Female has brownish-red and greyish-white eyes. Ground colour pale orange-yellow, with dark thoracic markings and dark broken dorsal line on abdomen. Young males similar to female. Easily distinguished from most other red skimmers by long dark legs, shortish abdomen, open wing venation and characteristic erect posture. Superficially similar Scarlet Basker (p. 319) is larger, with more extensive coloured patch at hindwing-base, and just two dark spots on abdomen, instead of dark dorsal line.

Habitat & Habits Occurs around ponds, reservoirs and marshes, especially near coasts and mangroves; also open country and reclaimed land near coasts. Perches conspicuously at tip of an exposed twig or topmost branches of shrubs and low trees. Migratory species that may fly high over the forest canopy but never in the understorey.

Presence in Singapore Recorded island-wide at the larger reservoirs, Sungei Buloh Wetland Reserve, Punggol, Bishan-Ang Mo Kio Park, the Singapore Botanic Gardens, Gardens by the Bay and Pulau Ubin.

Etymology The prefix *macro-* (Greek for 'long') refers to the spidery legs. *Diplax* is a common generic appellation for skimmers (see etymology of the genus *Brachydiplax*). Cora is the childhood name of Persephone, the Greek goddess of fertility.

Distribution Tropical Asia and Australasia reaching the Western Pacific; also the Middle East and eastern Africa.

National Conservation Status Least Concern; Widespread and Common.

IUCN Red List Status Least Concern.

Larva Squat with long, spidery legs. Semi-transparent and pelagic among water weeds.

Marcus Ng

Mature male, showing the long, mostly black legs and abdomen with a dark dorsal stripe.

Marcus Ng

Female, showing the typical perching position of the species.

Marcus Ng

Young male with paler colours.

Marcus Ng

Roosting female, showing the abdomen's dorsal markings.

Marcus Ng

The broken dorsal line on the abdomen is evident even in flight.

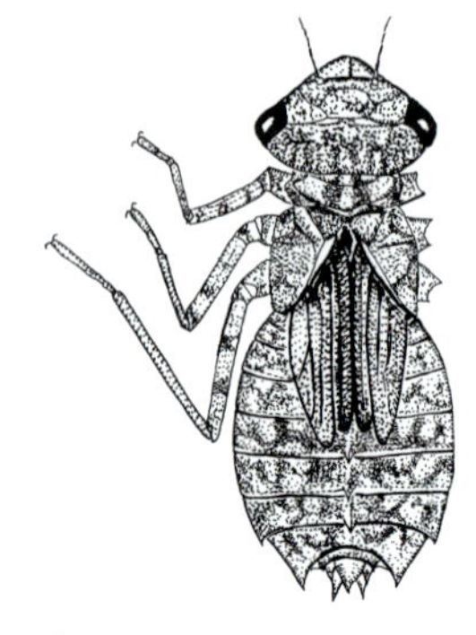

Albert G. Orr

Drawing of the larva.

SCARLET PYGMY *Nannophya pygmaea* Rambur, 1842

Size HWL: 12–13mm; TBL: 16–17mm

Description Tiny, bright red dragonfly. Smallest true dragonfly in Singapore (the smallest true dragonfly in Southeast Asia is *Nannophyopsis chalcosoma* of Borneo). Male has red and black eyes. Thorax has deep red and black markings; abdomen and appendages bright red. Immature males yellowish-brown, with paler eyes. Female has brown and light green eyes. Thorax has black and whitish markings. Abdomen stout, with transverse brown and whitish bands, possibly mimicking a bee or wasp. Wings of both sexes have broad amber tint at bases.

Habitat & Habits Found around open marshes, swamps and reservoir inlets in open country or near forests; also marshy areas of reclaimed land, especially areas with dense growth of low emergent vegetation, particularly sedges. Males active from around midday until late afternoon, defending small territories of about 1m^2, often perching in obelisk position. Females lurk in higher vegetation nearby until they are ready to come to water to mate.

Presence in Singapore Recorded in various marshy locations, including the Bukit Timah and Central Catchment Nature Reserves, the Western Catchment, Punggol, Tampines Eco-Green, Pulau Ubin and reclaimed land in Marina East and Tuas.

Etymology Generic epithet combines *nanos*, Greek for 'dwarf', and *phúō*, which means 'stature' or 'growth'.

Distribution Tropical and subtropical Asia and Australasia, from India, China, Korea and Japan, south to New Guinea.

National Conservation Status Least Concern; Widespread and Common.

IUCN Red List Status Least Concern.

Larva Very tiny (less than 10mm) and squat. Found in dense aquatic vegetation.

Marcus Ng

Fairly mature male in obelisk position.

Robin Ngiam

The bee-like mature female.

Marcus Ng

Maturing male that is developing its red colours.

Tang Hung Bun

Young male with yellowish rather than red hues.

Marcus Ng

Young female with paler colours.

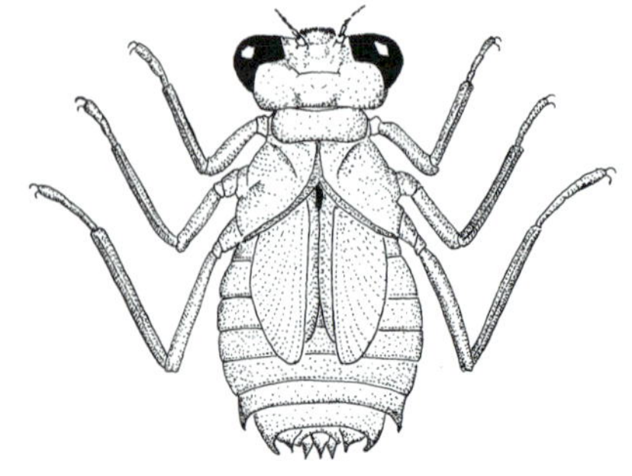

Albert G. Orr

Drawing of the tiny larva.

STRIPED GRENADIER *Nesoxenia lineata* (Selys, 1879)

Size HWL: 27–30mm; TBL: 34–37mm

Description Small-medium reddish dragonfly with spear-shaped abdomen and narrow, hyaline wings. Male has dark brown and bluish-green eyes. Thorax blackish with fairly straight and complete pale yellow stripes. Older individuals develop bluish-grey pruinescence that obscures thoracic markings. Abdomen dark and tapered at base, becoming broader at segments 6–8, which are also red with dark banding. Female similar in colour, with brown and light green eyes and straighter abdomen. Younger adults of both sexes have more extensive red on abdomen. Colours become partly obscured with age. Younger adults, with their redder abdomens, may be confused with the similarly sized and often sympatric Grenadier (p. 219), but that species has mottled thoracic markings and the abdomen is not tapered. Females may be confused with the Scarlet Grenadier (p. 251), but differ in size, thoracic markings and lack of sock-shaped anal loop.

Habitat & Habits Occurs in swampy forests and marshy areas with shaded pools. Not infrequently found at same sites as the Handsome Grenadier (p. 221), but prefers less prominent perches. Males guard small pools filled with leaf litter for long periods. Females and younger males may sometimes be encountered along trails near swampy forests.

Presence in Singapore Recorded in various forested locations such as the Central Catchment Nature Reserve, Windsor and Thomson Nature Parks, Rifle Range forest, Bukit Brown, Kranji Marshes, Admiralty Park and Pasir Ris. Described from a specimen collected by Wallace in Melaka.

Etymology Prefix *neso* means 'island' in Greek, while *xenos* means 'stranger'. The first known member of the genus was *N. mysis*, which was described based on specimens from the Solomon Islands.

Distribution Sundaland, Sulawesi and the Philippines.

National Conservation Status Least Concern; Widespread but Uncommon.

IUCN Red List Status Least Concern.

Larva To the best of the authors' knowledge, larva unknown for genus.

Marcus Ng

Mature male, showing the thoracic stripes (compare with the mottled markings of the Grenadier).

Marcus Ng

Striped Grenadier (left) and Handsome Grenadier sharing a perch. Note the differences in the eye colours, thoracic markings and abdomen shape.

Robin Ngiam

Dorsal view of the male, showing the spear-shaped abdomen.

Marcus Ng

Male, probably a younger individual, with fairly extensive red on the abdomen.

Marcus Ng

Young female with extensive red on the abdomen.

Marcus Ng

Older female with pruinescence on the base of the abdomen.

Rare Parasol *Neurothemis disparilis* Kirby, 1889

Size HWL: 23–25mm; TBL: 31–33mm

Description Very rare and poorly known dragonfly that is now known only from Borneo. Male similar to the Common Parasol (p. 263), but red in wings is reduced to basal quarter, ending well before nodus and rear margin. Abdomen has short lateral streaks. Female's wings hyaline but with strong yellowish basal tint on hindwings (more so than in female Common).

Habitat & Habits Probably occurs in swampy forests. Currently known only from a few locations in Borneo, mainly peat swamp forests and blackwater lakes.

Presence in Singapore Laidlaw collected specimens from Kuala Aring in Kelantan in 1899, and also examined specimens 'taken by Ridley in Singapore' around the turn of the twentieth century. Last putative local record was by Naoto Yokoi from Mandai in 1992, dubiously recorded as *N. intermedia*, a species not found in the Malay Peninsula but superficially similar to the Rare Parasol. Marcel Silvius, who encountered the species at Danau Sentarum National Park in West Kalimantan in 2019 and 2020, described the habitat: 'Danau Sentarum is a blackwater floodplain lake. There are many small islands in the floodplain, Pulau Ketenang and Pulau Sepandang being two of them. The lake floods in the wet season with up to 6 metres of water, and in the dry season it can be (almost) entirely dry. I found *Neurothemis disparilis* higher up on the islands in (non-forest) fern vegetation and near a small natural water pool on Pulau Sepandang, also surrounded by ferns.'

Etymology *Neuron* is Greek for 'nerve', and refers to male's densely veined wings, especially around hindwing-base. Suffix *-themis* is a common appellation for the names of skimmers. Specific epithet based on *dispar*, the Latin term for 'different'; most other species of red *Neurothemis* in the region have much more extensive red in the wings.

Distribution Known with certainty from Borneo (West Kalimantan). Old records mentioned by Laidlaw from Singapore and Kuala Aring may prove to be a different species, pending further study.

National Conservation Status Extinct.

IUCN Red List Status Data Deficient.

Larva Unknown, but should be similar to that of Common.

Reagan Villaneuva

Male specimen from Borneo.

Marcel Silvius

Male, showing the wing colouration, which is much less extensive compared to that of the Common Parasol. Photo taken in Borneo.

Marcel Silvius

Male photographed in a peat swamp in Borneo.

Marcel Silvius

Female photographed in Borneo.

Marcel Silvius

Male photographed in Borneo.

Dennis Farrell

Male Neurothemis intermedia *photographed in Thailand. Like the Rare Parasol, this species has reduced colour in the wings.*

Common Parasol *Neurothemis fluctuans* (Fabricius, 1793)

Size HWL: 22–25mm; TBL: 30–34mm

Description Small-medium dragonfly with richly coloured wings. Eyes brownish-red and brown; light brown and greenish in younger individuals. Thorax and abdomen of mature males exhibit various shades of brownish-red. Wings, except for tips, suffused with rich maroon, with colouration ending just before or at red pterostigmata (yellowish in immatures). In hindwing, outer border of colouration forms curve along rear margin. Hindwing-base very densely reticulated. Immature males light brown with thin abdominal streaks and paler wings. Younger males may be confused with the Common Amberwing (p. 229), which has more orange wing colouration, yellowish wing veins and more compact build. Female has brown and green eyes that turn brown and greyish with age, and yellow-brown body. Wings hyaline but may have small yellow basal tint, and reticulation less dense. Older females turn greyish and may have slightly darkened wing-tips. Andromorphs known.

Habitat & Habits Occurs at well-vegetated waterbodies. Found in diverse exposed habitats such as urban to rural ponds and streams, marshes, vegetated drains and open-canopy forest swamps. Both sexes often encountered foraging along open forest trails, forest edges and clearings, and grassland and parks some distance from water. Generally a low-flying species with a somewhat fluttery flight, and fairly easy to approach. Despite its abundance, mating seldom observed. Male attempts to seize a female without any courtship display, and if she is willing, the pair will remain in wheel for a few minutes at most; pairs in wheel are highly sensitive to disturbance, taking off for the canopy at the slightest movement. Preyed upon by other dragonflies such as the Common Flangetail, Variegated Green Skimmer and Common Redbolt (pp. 198, 278 and 292). Anecdotal observations indicate a wide variation in size within the local population, which should be further investigated.

Marcus Ng

Male with fairly extensive wing colours.

Presence in Singapore Widely distributed across the island. Probably Singapore's most common dragonfly species, along with the Wandering Glider (p. 282).

Etymology See previous species for etymology of the genus. Brauer coined the genus to replace *Polyneura* (Rambur, 1842), meaning 'many veined', a name that was found invalid as it had been earlier applied to a cicada. Specific epithet, Latin for 'fluctuating', refers to variable body colours that range from rusty-brown to brick-red.

Distribution Tropical Asia ranging from the Andaman and Nicobar Islands to Sundaland, Java and Palawan.

National Conservation Status Least Concern; Widespread and Common.

IUCN Red List Status Least Concern.

Larva Smallish and dark with banded, hairy long legs. Can be semi-pelagic or benthic. Found in shallow pools and tolerant of high water temperatures.

Marcus Ng

Smallish, lightly built individual in a common wing pose.

Marcus Ng

Male with reduced colour in the wings (red ending before the pterostigmata).

Marcus Ng

Young male with pale wing and body colours.

Marcus Ng

Female, showing the mostly hyaline wings with more open venation and abdominal markings (compare with the Common Amberwing).

Marcus Ng

Older female with duller colours and slightly darkened wing-tips.

Marcus Ng

Andromorph female.

Meena Vathyam

Pair in wheel.

Robin Ngiam

The larva showing its hairy legs.

RIVERHAWK *Onychothemis testacea* Laidlaw, 1902

Size HWL: 37–40mm; TBL: 47–50mm

Description Fairly large, powerfully built dragonfly with distinctive markings. Wings hyaline. Eyes dark green; brown-greyish in younger individuals. Thorax and abdomen deep metallic greenish to blackish-blue, with pale yellow lines on thorax and basal abdominal segments, and irregular orange-yellow bands on segments 4–8. Legs (especially hindlegs) long, with particularly large, sparse spines on tibia. Sexes similar.

Habitat & Habits Found around clear, moderate to fast-flowing forest streams. Breeds in riffles. Young adults may forage at forest edges some distance from water. Perches in prominent positions, but often hard to approach. Appears to be specialist predator of butterflies and other large insects, capturing such prey with aid of robust tibial spines.

Presence in Singapore Recorded sporadically in the Central Catchment Nature Reserve.

Etymology *Onycho-* (Greek for 'talon' or 'claw') refers to a peculiar feature of legs. Brauer, who coined the genus in 1868, wrote: 'Legs strong and long, tibiae curved, with a few long robust spines well spaced out, which are curved and as strong as the claws. Claws without a tooth, only those of the last pair of legs with a little notch outside of the middle.' Specific epithet means 'brick-red'. Laidlaw did not explain the etymology, but noted 'a reddish-brown spot on either side' of the labrum.

Distribution South Asia, parts of China, Taiwan and mainland Southeast Asia.

National Conservation Status Endangered; Restricted and Very Rare.

IUCN Red List Status Least Concern.

Larva Flattened, with strong legs and claws. Inhabits clear forest streams and riffles.

Marcus Ng

Male at a forest edge near Upper Seletar Reservoir.

Richard Ong

Male feeding on a butterfly (Ypthima sp.) near MacRitchie Reservoir.

Marcus Ng

Young male (note the brownish eyes) in Gopeng, Malaysia.

Tang Hung Bun

Dorsal view of a female photographed in Belumut, Malaysia.

Antonio Giudici.

Female in Thailand feeding on a flatid planthopper.

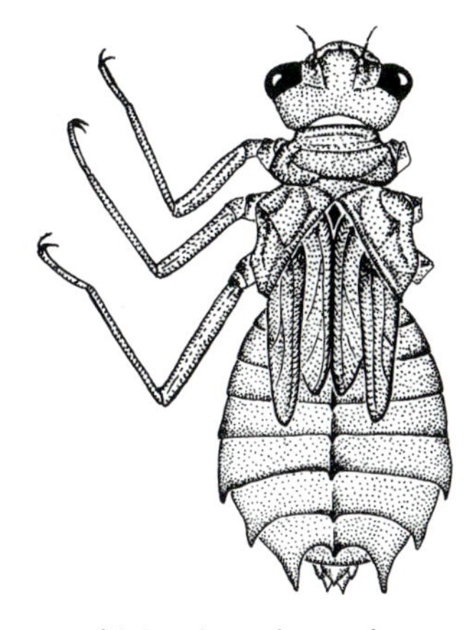

Albert G. Orr

Drawing of the larva showing the strong claws.

BLUE SENTINEL *Orchithemis pruinans* (Selys, 1878)

Size HWL: 26–27mm; TBL: 32–33mm

Description Small-medium blue dragonfly with narrow, hyaline wings. Eyes dark brown and greyish-blue. Male's thorax and abdomen dark blue-black, with bluish-white pruinescence spanning segments 3–4. Female brownish or greyish-blue, lacking abdominal basal pruinescence. May be confused with blue form of the Variable Sentinel (p. 270), but Blue is slightly larger, with a longer and thinner abdomen. Also, in the blue form of Variable, pale section of male's abdomen covers just segment 3.

Habitat & Habits Found around shaded pools and small rivulets in swampy forests, which are guarded by male. Prefers well-shaded spots, and seldom ventures to forest edges, unlike Variable.

Presence in Singapore Restricted to Nee Soon Swamp Forest.

Etymology Prefix *orchi-* (Greek for 'testicle') refers to shape of male's secondary genitalia. Specific epithet comes from *pruinosus*, meaning 'waxy'.

Distribution Singapore, Peninsular Malaysia, southern Thailand, Sumatra and Borneo.

National Conservation Status Critically Endangered; Restricted and Rare.

IUCN Red List Status Least Concern.

Larva Typical libellulid in appearance. Dark brown, oviform abdomen. Found among pool bottom debris.

Marcus Ng

Dorsal view of the male, showing the blue on abdominal segments 3-4.

Marcus Ng

Lateral view of the male. This individual has a Forcipomyia biting midge feeding from its abdomen.

Marcus Ng

Close-up of the male's head, showing the eye and facial colours.

Eugene Tay

Pair in wheel.

Variable Sentinel *Orchithemis pulcherrima* Brauer, 1878

Size HWL: 21–23mm; TBL: 29–32mm

Description Small-medium dragonfly with broad, tapering abdomen and very different colour morphs. Eyes dark brown and green, with dark edging. Wings narrow and hyaline, with rounded tips and lacking sock-shaped anal loop. Males occur in two forms: red and blue. Red morph has pale yellow thorax and largely bright red abdomen, with segment 10 and appendages black. Blue morph, which is less abundant but may be locally common, has bluish-black abdomen, with conspicuous bluish-white pruinescence on segment 3. Females also occur in two primary morphs. One has yellow-brown thorax (with dark dorsum) and bluish-grey abdomen. The other, less common form, resembles blue morph male, but bluish-white pruinescence on abdomen often extends into segments 2 and 4. There is also a rare red form, which could be regarded as an andromorph. Other supposed colour morphs are probably individuals at varying stages of maturity. Young adults of both sexes uniformly yellow-brown.

Habitat & Habits Found at swampy forests and forest streams. Along with the Treehugger (p. 316), probably the most common true dragonfly in deep forest. May also be abundant in disturbed forest habitats. Male guards small rivulets or pools in shaded, swampy forests. Rivals hover before each other in a lively agonistic display with their abdomens raised almost vertically, ascending in the air until one breaks off and is chased away. Both sexes, especially immatures, may also be seen foraging at forest clearings or edges. The blue form, with its highly contrasting colours, may be mistaken for a wasp when in flight.

Presence in Singapore Recorded in various locations, including the Central Catchment and Bukit Timah Nature Reserves and forested nature parks, the Western Catchment, Bukit Batok, Bukit Brown, Admiralty Park, Pulau Ubin and Pulau Tekong.

Etymology Specific epithet derived from *pulcherrimus*, Latin for 'most beautiful', perhaps referring to dragonfly's varied and striking colours.

Distribution Sundaland, Vietnam and the Philippines.

National Conservation Status Least Concern; Widespread and Common.

IUCN Red List Status Least Concern.

Larva Similar in appearance to larvae of the Blue Sentinel (p. 268).

Marcus Ng

Dorsolateral view of a typical red form male, with a pale yellow thorax and mostly red abdomen.

Marcus Ng

Blue morph male with pruinescence on abdominal segment 3.

Marcus Ng

Young blue form male with pale abdominal colours.

Marcus Ng

Young male. The young female has similar colours.

Marcus Ng

Typical mature female.

Marcus Ng

Blue morph female with pruinescence restricted to abdominal segment 3.

Marcus Ng

Blue morph female with pruinescence on abdominal segments 2–4.

Marcus Ng

Red female. Note the ventral flaps on the abdomen.

Robin Ngiam

The light brown larva with squarish head.

Spine-tufted Skimmer *Orthetrum chrysis* (Selys, 1891)

Size HWL: 31–34mm; TBL: 41–48mm

Description Medium-sized, mostly red dragonfly. Male has bluish-grey eyes and 'dirty' brown thorax. Abdomen mostly bright red and swollen basally, with spine-like tuft of setae below segment 2. Appendages red. Wings hyaline except for slight yellow tint at hindwing-base. Female has brown-grey eyes, dull reddish-brown abdomen, and small dark ventral flaps beneath segment 8. May be confused with the Scarlet Skimmer (p. 280), which may occur with this species, but male Scarlet has orange-brown thorax and more extensive hindwing tint, while female is lighter yellow-brown.

Habitat & Habits Found around ponds, streams and marshes with ample surrounding vegetation, including back ends of mangroves. Occurs in disturbed and semi-open urban habitats as well as in forests. The only local *Orthetrum* species that is common in closed forest streams. Males fairly conspicuous by the water, where they maintain territories and chase off other red dragonflies. Females scarcer, usually encountered while in wheel or ovipositing near weedy banks with male in attendance.

Presence in Singapore Recorded in many locations, including the Bukit Timah, Labrador and Central Catchment Nature Reserves and adjacent nature parks, Sungei Buloh Wetland Reserve, Admiralty Park, Toa Payoh Town Park, the Rail Corridor, Tampines Eco-Green, Singapore Botanic Gardens, Kent Ridge Park, Bukit Brown and Pulau Ubin.

Etymology *Orthetrum* is derived from *orthos* (Greek for 'straight'), referring to shape of abdomen. Edward Newman, who coined the genus in 1833, separated the formerly catch-all genus *Libellula* into, among others, *Sympetrum*, in which abdomen is 'laterally compressed', and *Orthetrum*, which has a

Marcus Ng

Dorsal view of the male, showing the small hindwing-base tint and complete distal antenodal crossveins.

'laterally parallel' abdomen. Chrysis was a priestess of the Greek goddess Hera, who inadvertently caused a fire that destroyed her temple – not an inappropriate name for a dragonfly with such fiery colours.

Distribution Widespread in tropical Asia.

National Conservation Status Least Concern; Widespread and Common.

IUCN Red List Status Least Concern.

Larva Smallish, rectangular head and bean-like eyes. Very hairy, and often found covered with bottom detritus.

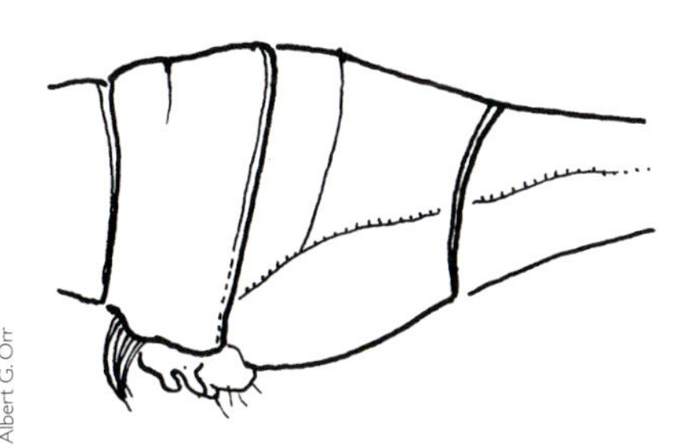

Albert G. Orr

Drawing showing the tuft of setae beneath abdominal segment 2 of the male.

Marcus Ng

A male, showing the blue-grey eyes and dark brown thorax.

Marcus Ng

The reddish-brown female, showing its dark ventral flaps.

Marcus Ng

Older female with duller colours.

Marcus Ng

Pair in wheel.

Robin Ngiam

The rather hairy larva.

Common Blue Skimmer *Orthetrum glaucum* (Brauer, 1865)

Size HWL: 32–35mm; TBL: 41–46mm

Description Medium-sized blue dragonfly that often perches low or on the ground. Male has greenish-blue eyes and dark blue thorax. Abdomen lighter blue, with segments 9–10 and appendages dark. Wings hyaline, with small dark tint at hindwing-base. Female has dull blue eyes; eyes brownish in younger individuals. Thorax yellow-brown with two thick dark lateral stripes. Abdomen mostly light brown. Young male similar to female. Older females develop powdery-blue pruinescence that covers most of body.

Habitat & Habits Found at drains, ponds and streams in open, disturbed habitats or near forests. Males often perch at sides of concrete ditches at forest edges or on sunny trails, where they actively pursue rivals and females from about midday. Also fond of perching on rocks or bare ground.

Presence in Singapore Recorded in many forested as well as semi-urban to urban locations, including the Bukit Timah and Central Catchment Nature Reserves and adjacent nature parks, Bukit Batok Nature Park, Kent Ridge Park, Pearl's Hill Park, Bishan-Ang Mo Kio Park, Fort Canning Park and Jurong Central Park.

Etymology Specific epithet comes from *glaukos*, Greek for 'greyish-blue'.

Distribution Widespread in tropical and subtropical Asia.

National Conservation Status Least Concern; Widespread and Common.

IUCN Red List Status Least Concern.

Larva Typical of genus.

Marcus Ng

Mature male, perched on the concrete side of a drain.

Marcus Ng

Young female with no pruinescence. Note the two dark brown stripes on the side of the thorax (compare with the Slender Blue Skimmer, p. 276, which has just one side stripe).

Marcus Ng

Very young male with colours that resemble the female's.

Marcus Ng

Old female with pruinescence covering her body.

Marcus Ng

Maturing male with developing blue pruinescence.

Marcus Ng

A pair in wheel.

Slender Blue Skimmer *Orthetrum luzonicum* (Brauer, 1868)

Size HWL: 30–32mm; TBL: 40–42mm

Description Medium-sized light blue dragonfly with hyaline wings. More lightly built than the Common Blue Skimmer (p. 274). Male has azure-blue eyes. Thorax and abdomen covered by light blue pruinescence. Thorax of younger males brownish-yellow with one thick dark stripe on each side. Females have blue-grey eyes, and light brown body with dark abdominal markings. Older females develop bluish pruinescence. Distinguished from Common by more slender abdomen, eye colours, fully hyaline wings and thoracic markings (only one dark lateral stripe; Common has two).

Habitat & Habits Elsewhere in region, associated with submontane or montane areas, but locally occurs at streams and marshes in open country or forest edges. Less fond of lentic habitats compared to other local *Orthetrum* species. Favours streams with grassy banks flowing through semi-open, marshy areas.

Presence in Singapore Recorded in various locations such as the Bukit Timah and Central Catchment Nature Reserves, Windsor and Bukit Batok Nature Parks, Admiralty Park, the Rail Corridor, Holland Woods and Bukit Brown. Least common member of its genus locally.

Etymology Species was described based on specimens obtained from Luzon in the Philippines, hence the specific epithet.

Distribution Widespread in tropical Asia, north to Japan.

National Conservation Status Least Concern; Widespread and Common.

IUCN Red List Status Least Concern.

Larva Typical of genus.

Marcus Ng

Mature male. Note the sky-blue eyes and slender abdomen.

Marcus Ng

Lateral view of a young male, showing the thoracic markings.

Marcus Ng

Dorsal view of a young male.

Marcus Ng

Dorsal view of a young female. Note the single brown stripe on the side of the thorax and slender abdomen.

Marcus Ng

Older female, with pruinescence developing on the abdomen.

Variegated Green Skimmer *Orthetrum sabina*
(Drury, 1773)

Size HWL: 32–35mm; TBL: 47–52mm

Description Fairly large but slender, green and black dragonfly. Both sexes have green eyes and yellowish-green thorax with variegated black markings. Abdomen pale and swollen basally, more so in female than in male, with prominent black banding. Terminal segments slightly expanded and black. Appendages white. Colours and shape of abdomen may cause confusion with clubtails (Gomphidae), which can be told apart by their well-separated eyes, and the acute rear margin of the hindwing in males. Young males and females distinguished from female Blue Percher (p. 243) by larger size and more heavily marked thorax.

Habitat & Habits Favours still and slow-flowing waterbodies in fairly open habitats, including drains, ponds, reservoirs and marshes. Also fairly common in forest edges, mangroves and coastal wetlands. Often strays into urban gardens, fields and compounds. Can be inconspicuous when perched amid low vegetation by water. Voracious predator that often preys on other odonates, including damselflies and other true dragonflies, as well as butterflies.

Presence in Singapore Occurs in nature reserves and parks all over the main island and offshore islands, including Pulau Ubin, Sentosa and Pulau Semakau.

Etymology Species named after the Sabines, an ancient Italic people with a reputation for war-like or aggressive behaviour.

Distribution Old World tropics from the Mediterranean and East Africa to East Asia and Australasia.

National Conservation Status Least Concern; Widespread and Common.

IUCN Red List Status Least Concern.

Larva Typical of genus. Overall hairy. Small, bean-like eyes. Found in muddy or leaf litter bottoms. Apparently salt tolerant.

Marcus Ng

Male, showing the greatly expanded base of the abdomen and long, pale anal appendages.

Marcus Ng

Female, showing the abdominal cerci.

Marcus Ng

Pair in wheel.

SCARLET SKIMMER *Orthetrum testaceum* (Burmeister, 1839)

Size HWL: 34–38mm; TBL: 43–48mm

Description Fairly large, vermillion red dragonfly. Male has light brown-grey eyes and orange-brown thorax. Abdomen vermilion red. Hindwing has amber patch at base. Female has brown eyes that turn greyish with age, and mostly unmarked yellow-brown body – lighter in colour than that of the female Spine-tufted Skimmer (p. 272) – which becomes duller (greyish) with age. Superficially similar male Spine-tufted has 'dirtier' thorax, brighter red abdomen and smaller hindwing tint. Distinguished from the slightly smaller Common Scarlet (p. 239) by unmarked abdomen and from the Common Redbolt (p. 292) by eye shape and colours. All species of *Orthetrum* also have complete distal antenodal crossvein in forewing (crossvein just before nodus spans two cells), whereas in *Crocothemis* and *Rhodothemis* species distal antenodal is incomplete, spanning just one cell.

Habitat & Habits Occurs at drains, ponds, gardens and forest edges. Common in disturbed, open habitats, including urban parks; seldom encountered in deep forests.

Presence in Singapore Recorded island-wide in the Central Catchment and Bukit Timah Nature Reserves, Sungei Buloh Wetland Reserve, the Western Catchment, and various nature and town parks.

Etymology Specific epithet (Greek for 'brick-coloured') refers to reddish ground colour.

Distribution Widespread from India to Southeast Asia, Sulawesi and New Guinea.

National Conservation Status Least Concern; Widespread and Common.

IUCN Red List Status Least Concern.

Larva Typical of genus.

Marcus Ng

Dorsal view of the male, showing the orange-red thorax and tint at the hindwing-bases.

Scarlet Skimmer (right) sharing a perch with a Spine-tufted Skimmer (left). Note the differences in eye and thoracic colours and the extent of colour in the hindwing-base.

Lateral view of a male.

Younger female with typical brown-yellow hues.

Grey-brown mature female. The colours are paler compared to those of the female Spine-tufted Skimmer.

Pair in wheel. Photo taken in Sabah, Malaysia.

WANDERING GLIDER *Pantala flavescens* (Fabricius, 1798)

Size HWL: 39–41mm; TBL: 45–47mm

Description Large, yellowish-brown dragonfly that is usually on the wing. Eyes orange-brown and bluish-grey. Thorax light brown and rather hairy. Abdomen straight and tapered, sandy yellow to rich orange-brown (some males have reddish tinge), with series of broken dark dorsal markings that become more well defined towards tip. Appendages long, brown and tipped with black. Wings broad and hyaline, except for small dark mark at tip of hindwing. Female similar to male, with long cerci.

Habitat & Habits Common in open and urban environments; sometimes forms feeding swarms above wayside trees, canals and grassy patches, even in the city centre. Highly aerial, often gliding and soaring for long periods, but seeks perches in low vegetation during overcast weather or before rain. Perches in hanging position with abdomen pointing downwards, at times in small groups. Breeds in stagnant and slow-flowing waterbodies. Early colonizer of newly cleared land, where temporary pools may form after heavy rains. In urban settings, may follow cars and attempt to oviposit on their shiny hoods or roofs, perhaps mistaking these surfaces for water.

Presence in Singapore Occurs throughout the island in almost every habitat except deep forest, though may forage above the canopy. Quite often seen flying over the Singapore Strait and offshore islands and reefs.

Etymology *Pantala* derives from a Greek term meaning 'to wander without hope of rest', describing the species' vagrant habits. Specific epithet means 'yellowish' in Latin.

Distribution Worldwide in tropical and temperate regions (but rare in Europe), including oceanic islands. Cosmopolitan and migratory species that wanders to distant oceanic islands and mountain tops, hence also called the Globe Skimmer. Mascot of the Worldwide Dragonfly Association.

National Conservation Status Least Concern; Widespread and Common.

IUCN Red List Status Least Concern.

Larva Similar to larvae of *Hydrobasileus* species, but more slender and abdomen less spiny. Found in many still and slow-flowing waterbodies, including temporary pools. Early instar drought resistant. The time from egg to adult may take just six weeks.

Marcus Ng

Male flying over a field in a housing estate.

Marcus Ng

Dorsal view of a perched male, showing the abdominal markings, broad hindwings and small dark mark at each wing-tip.

Marcus Ng

During flight the long hindlegs are folded beneath the synthorax.

Marcus Ng

Lateral view of a male, showing the long upper anal appendages.

Marcus Ng

Dorsal view of a female, showing the long cerci.

Marcus Ng

Lateral view of a female. Note the absence of lower anal appendages.

Mangrove Marshal *Pornothemis starrei* Lieftinck, 1948

Size HWL: 29–31mm; TBL: 41–44mm

Description Medium-sized, darkish dragonfly with very slender, arched abdomen. Wings hyaline and narrow; hindwing without sock-shaped anal loop. Male has dull blue eyes. Thorax and abdomen almost entirely bluish-black, with bluish-white pruinescence at wing-bases and base of abdomen. Dorsum of synthorax has patch of light green spots, which may be obscured by pruinescence. Abdominal segments 1–3 inflated, thereafter very narrow and curved in profile. Female has pale green eyes, olive-green thorax, and markedly thicker, blackish abdomen with ventral flaps beneath segment 8; may be confused with the Variegated Green Skimmer (p. 278) from a distance. Younger male similar to female. Local specimens previously wrongly identified as *P. serrata* by Laidlaw and Murphy.

Habitat & Habits Restricted to coastal mangroves and nypa swamps, seldom straying far from this specific habitat. Males perch on low vegetation or aerial roots of mangrove trees in semi-open mudflats or tidal creeks. Females oviposit in small tidal pools by repeatedly flicking the abdomen at the surface, probably to fling batches of wetted eggs at muddy banks.

Presence in Singapore Recorded in several mangrove sites, including Sungei Buloh Wetland Reserve, Admiralty Park, Mandai Mangroves, Lim Chu Kang, Coney Island, Pulau Ubin and Pulau Tekong.

Etymology In Greek, *pornos* means 'fornicator'. Leopold Krüger, who coined the genus in 1902, described the male's secondary genitalia as *vortretend* ('protruding' in German). Species named after Captain J. J. van der Starre, who collected the type specimen in Sumatra in 1938.

Distribution Singapore, Sumatra and Borneo. Known from only eight locations in region.

National Conservation Status Near Threatened; Widespread but Uncommon.

IUCN Red List Status Near Threatened.

Larva Unknown for genus. Discovery of larva would be of great scientific interest as it breeds in very saline conditions, which is unusual among dragonflies.

Marcus Ng

Dorsolateral view of a male, showing the green spots on the synthorax and thin, arched abdomen.

Marcus Ng

Lateral view of a mature male.

Marcus Ng

Young male with pale colours.

Yue Teng Lee

A female. Note the thicker abdomen with ventral flaps.

Laurence Leong

Pair in wheel.

Marcus Ng

Female preparing to oviposit in an intertidal pool within a mangrove swamp.

COMMON CHASER *Potamarcher congener* (Rambur, 1842)

Size HWL: 31–34mm; TBL: 43–45mm

Description Medium-sized dragonfly with contrasting blue and orange colours. Wings hyaline and narrow, but with sock-shaped anal loop. Eyes reddish-brown and greyish-white. Male's thorax and basal segments of abdomen covered by powdery-blue pruinescence. Remaining segments dark with orange-yellow lateral streaks, except for segments 9–10, which are black. Female's thorax has light yellow and dark brown stripes that are thicker on dorsum. Abdomen dull orange with dark banding; prominent ventral flaps beneath segment 8, distinguishing it from the female Scarlet Grenadier (p. 251). Colours become obscured with age. Younger male has similar thoracic markings to female.

Habitat & Habits Found at open-country ponds, disturbed semi-urban habitats and urban habitats with standing or slow-flowing water. Seldom near forests. In parts of India, known to roost in large aggregations of about 100 individuals per site on certain trees during the dry season.

Presence in Singapore Recorded island-wide in various locations, including Sungei Buloh Wetland Reserve, Pasir Ris Park, Sengkang Riverside Park, Tiong Bahru Park, Gardens by the Bay, East Coast Park, Kranji, Jurong Lake Gardens and HortPark.

Etymology Generic epithet combines *potamos* (Greek for 'river') and *arkhós* ('chief' or 'leader'). *Congener* means 'of the same kind'. Rambur may have used this specific epithet due to perceived similarities with *Libellula* (later *Potamarcha*) *obscura*, a taxon he had described earlier in the same publication, which was later synonymized with *congener*.

Distribution Tropical Asia and Australasia.

National Conservation Status Least Concern; Widespread and Common.

IUCN Red List Status Least Concern.

Larva Typical libellulid in appearance. Very hairy.

Marcus Ng

A mature male. Note the blue thorax and abdomen-base, and orange-yellow streaks on the abdomen.

Marcus Ng

Mature female, showing the thoracic markings and ventral flaps beneath abdominal segment 8 (absent in the female Scarlet Grenadier).

Marcus Ng

Young male with thoracic markings similar to the female's.

Marcus Ng

Older female, with thoracic markings obscured by pruinescence.

Marcus Ng

Dorsal view of the female, showing the orange abdominal streaks.

Choong Chee Yen

The hairy larva.

BANDED SKIMMER *Pseudothemis jorina* Förster, 1904

Size HWL: 31–33mm; TBL: 37–38mm

Description Medium-sized, highly aerial dragonfly with unmistakable pied patterns. Wings hyaline except for small dark brown streaks at hindwing-base. Male has dark brown and blackish eyes. Body blackish except for white frons and bright creamy-white band (yellowish in immature males) near base of abdomen. The only other true dragonfly with a similar pattern is the blue morph of the Variable Sentinel (p. 270), which is much smaller and restricted to swampy forests. Female has brown and grey-green eyes, and black and yellowish body markings. Dorsum of basal abdominal segments bright yellow.

Habitat & Habits Occurs at ponds, lakes and reservoirs in parks, open country and forest edges. Prefers open waters with not too much vegetation. Highly aerial species, especially towards noon, when males actively patrol the water's edge, stopping to briefly hover at particular spots before moving on and returning later as part of a circuit. They may rest briefly on floating or emergent vegetation. During overcast days and later in the day, males may be found clinging to reeds or tree branches. Sometimes joins feeding swarms dominated by other species such as the Yellow-barred Flutterer and Wandering Glider (pp. 298 and 282). Females less often seen, but may be encountered on vegetation close to water. Copulation takes place in mid-air and female oviposits on floating twigs while male hovers nearby.

Presence in Singapore Recorded in various locations such as the Bukit Timah and Central Catchment Nature Reserves and adjacent nature parks, Sungei Buloh Wetland Reserve, Kent Ridge Park, Jurong Lake Gardens, Pasir Ris, Singapore Botanic Gardens, the Istana, Punggol Park and Pulau Ubin.

Etymology *Pseudo* means a 'lie' or something false, while *themis* is a common appellation for libellulid genera. Kirby coined the genus in 1889 for *P. zonata* (Pied Skimmer), a very similar species from East Asia, which was earlier known as *Libellula zonata*. Specific epithet derives from *joris*, Greek for 'watchful' or 'alert'.

Marcus Ng

Lateral view of a perched male, showing the unique markings.

Robin Ngiam

Dorsal view of a male, showing the wing-base markings.

Distribution Mainland Southeast Asia and Borneo.

National Conservation Status Least Concern; Widespread but Uncommon.

IUCN Red List Status Least Concern.

Larva To the best of the authors' knowledge, larva undescribed, but should be similar to that of *P. zonata*, with wide, oval abdomen and strongly dark-banded legs.

Erland Refling Nielsen

A male as typically encountered, patrolling the edges of a pond. Photo taken in Kepong, Malaysia.

Tang Hung Bun

The dorsal marking on the female's abdomen is bright yellow.

Marcus Ng

Female photographed in Malaysia.

Bill Ho (AFCD)

Larva of the closely related Pseudothemis zonata. *Photo taken in Hong Kong.*

Mangrove Dwarf *Raphismia bispina* (Hagen, 1867)

Size HWL: 21–23mm; TBL: 26–29mm

Description Small dragonfly with blue males and 'tiger-striped' females, associated with mangroves. Eyes brown and yellowish-green. Wings hyaline and fairly narrow, with 'open-toed' anal loop. Male's thorax and most of abdomen slate-blue. Abdominal segments 7–10 and appendages black. Underside of synthorax (between leg-bases) has two short spines. Female dark metallic blue with mottled yellow markings on thorax and yellow streaks on abdomen, which fade with age. Male may be confused with the Blue Dasher (p. 223), but has 'cleaner' appearance, without brown on thorax, and fully hyaline wings. Also distinguished by its smaller size, habitat preference and number of antenodal crossveins (around 10 versus seven in the Blue Dasher). Distinguished from the similarly sized Pond Adjutant (p. 217) by fully hyaline wings, denser venation and 'open-toed' anal loop.

Habitat & Habits Restricted to mangroves. Males guard small, brackish pools subject to tidal influence. During sunny days, males take part in agonistic flight displays, hovering actively before each other and slowly ascending until one breaks off – with the other giving chase. Both rivals usually return to their original perches eventually. Females may also be seen along trails near mangrove swamps and creeks.

Presence in Singapore Locally recorded in various mangrove habitats, including Sungei Buloh Wetland Reserve, Admiralty, Marsiling and Pasir Ris Parks, Berlayar Creek, Mandai Mangroves, Pulau Semakau, Pulau Ubin and Pulau Tekong.

Etymology Generic epithet (from *raphis*, 'needle' in Greek) refers to two spines on underside of metasternum (rearmost ventral plate of thorax). Kirby coined the genus in 1889 for *Diplax bispina* in recognition of this unique anatomical feature. Specific epithet also refers to these spines. A closely related species found only in Bornean peat swamps, *R. inermis*, lacks the spines (*inermis* means 'unarmed').

Distribution Sundaland, the Philippines and Australasia.

National Conservation Status Near Threatened; Widespread but Uncommon.

IUCN Red List Status Least Concern.

Larva Unknown for genus. Of scientific interest due to its saline habitat.

Marcus Ng

Lateral view of the male. Note the thorax, which lacks brown and is less boxy compared to that of the Blue Dasher.

Marcus Ng

Dorsal view of the male. Note the hyaline wings and greater number of antenodals (10) compared to the Blue Dasher (7).

Robin Ngiam

Close-up of the thorax showing two short spines.

Tang Hung Bun

Young female with its striking markings.

Marcus Ng

Older female with obscured markings.

Marcus Ng

Pair in wheel.

Common Redbolt *Rhodothemis rufa* (Rambur, 1842)

Size HWL: 33–35mm; TBL: 41–44mm

Description Medium-sized red dragonfly. Male has deep red eyes, without pale edging of the Common Scarlet (p. 239). Thorax deep red, with paler red line running down dorsum, extending into abdomen. Abdomen without dark spots or markings. Wings hyaline, with light brown patch at hindwing-base. Female has rich brown eyes and thorax. Pale yellowish-white line runs down dorsum of thorax, extending to abdominal segments 3–4. Abdomen mostly light brown. Wing-base tint lighter and smaller. Immature male similar to female. Distinguished from the Common Scarlet by lack of dark dorsal stripe on abdomen, darker legs, and eyes that touch at a single point when viewed from above. Distinguished from red *Orthetrum* species by incomplete distal antenodal crossvein in forewing (complete in *Orthetrum*). Hindlegs also relatively longer and more spiny compared to other red skimmers.

Habitat & Habits Occurs at well-vegetated drains, ponds, marshes and reservoir edges. Sun-loving species absent from deep forests. Fairly wary insect that may perch low but readily flees to higher perch when approached. Known to prey on other dragonflies such as the Common Parasol (p. 263).

Presence in Singapore Recorded island-wide in natural to urban habitats, including the Bukit Timah and Central Catchment Nature Reserves and adjacent nature parks, Sungei Buloh Wetland Reserve, Kranji Marshes, Kent Ridge Park, Toa Payoh Town Park, Singapore Botanic Gardens, Bishan-Ang Mo Kio Park, Pasir Ris, Sengkang Riverside Park and Pulau Ubin.

Etymology Generic epithet combines the Greek for 'rosy' (*rhodon*) and *themis*, a common appellation for libellulid dragonflies. Specific epithet means 'rufous' (reddish-brown) in Latin.

Distribution Widespread in subtropical and tropical Asia, from India to New Guinea and the Solomon Islands.

National Conservation Status Least Concern; Widespread and Common.

IUCN Red List Status Least Concern.

Larva Very hairy and roundish abdomen. Bulbous eyes. Found among aquatic vegetation.

Marcus Ng

Dorsal view of a male, showing the eyes, which meet at a single point, and the incomplete distal antenodal crossveins in the forewings.

Marcus Ng

Dorsolateral view of a male. Note the lack of a dark abdominal stripe (compare with the Common Scarlet).

Marcus Ng

Dorsal view of a female. Note the pale dorsal stripe that reaches abdominal segment 4.

Tang Hung Bun

Pair in wheel.

Marcus Ng

Male feeding on a Common Parasol.

Robin Ngiam

The very hairy larva.

Small Bronze Flutterer *Rhyothemis fulgens* Kirby, 1889

Size HWL: 21–23mm; TBL: 25–27mm

Description Small dark dragonfly with entirely opaque, bronzey wings. Eyes dark brown. Abdomen dark and short, less than twice width of hindwing. Hindwing-base greatly expanded. Wings entirely bronze-brown with purplish reflex, but without variegated light and dark markings, unlike in the Bronze Flutterer (p. 296), which also has comparatively longer abdomen. Sexes similar.

Habitat & Habits Found at small streams in swampy forests and clear rocky streams at forest edges. Where present, males active on sunny days, perching with wings depressed or at an angle against the sun. Flight fluttery. Marcel Silvius encountered the species in Brunei at the 'edge of a wet kerangas forest in grassy, herbaceous 1m-high vegetation with small bushes, on burned-over kerangas'.

Presence in Singapore Originally identified as *R. pygmaea* (Brauer, 1867), of which *R. fulgens* was regarded as a junior synonym. Now recognized as a distinct species, with *R. pygmaea* occurring east of the Wallace Line (east of Borneo and Bali), and Small Bronze in the rest of Southeast Asia. Wallace collected a male from Singapore in 1854.

Etymology Generic epithet thought to derive from rhyolite, a volcanic rock with varied and irregular colour patterns that resemble wing markings of this genus. *Fulgens* is Latin for 'shining' or 'bright'.

Distribution Singapore (formerly), Peninsular Malaysia, Sumatra and Borneo.

National Conservation Status Extinct.

IUCN Red List Status Near Threatened.

Larva Unknown, but should be typical of genus (see entry for the Yellow-barred Flutterer, p. 298).

Choong Chee Yen

Male photographed in Pahang, Malaysia.

Marcel Silvius

Female photographed in Brunei.

Robin Ngiam

Lateral view of the male's anal appendages.

Marcel Silvius

Female photographed in kerangas forest in Brunei.

BRONZE FLUTTERER *Rhyothemis obsolescens* Kirby, 1889

Size HWL: 21–23mm; TBL: 24–28mm

Description Smallish dark dragonfly with brilliant wing colours. Eyes brownish-red and dark brown; thorax and abdomen dark brown to blackish. Abdomen length more than twice width of hindwing; compare with relatively shorter abdomen of the Small Bronze Flutterer (p. 294). Hindwing-base well expanded, though less so than in Small Bronze. Wings entirely opaque and metallic bronze, with iridescent light and dark markings and magenta reflex in good light. May be mistaken for the Common Parasol (p. 263) from afar, but its smaller size, shiny wings and different basking habits are evident at close range. Female similar to male, but usually with hyaline patches near wing-tips.

Habitat & Habits Favours well-vegetated open marshes, ponds, reservoir inlets and edges of swampy forests. Both sexes, but especially females, may also bask some distance from water on hilltops, tree canopies and masses of fallen branches in forest edges and clearings. May form feeding swarms or soar over the canopy, sometimes in mixed flocks with other species such as the Wandering Glider (p. 282). Perches with wings depressed or held at an angle. Often basks at tip of a twig or reed, its wings gleaming as they tilt and turn to catch the sun. Typical flight fluttery and slow. Pairs engage in delicate courtship dance, with each individual taking turns to hover slightly above and in front of the other. This 'waltz' is performed repeatedly before, as well as after, copulation.

Presence in Singapore Recorded in forests and well-vegetated locations near forests, including the Bukit Timah and Central Catchment Nature Reserves, Windsor Nature Park, the Singapore Botanic Gardens, the Istana, the Western Catchment, Kent Ridge Park and Pulau Ubin.

Etymology In Latin, *obsolescens* means 'passing out of use'. Specific epithet may be a reference to obsolete status of bronze.

Distribution Sundaland, mainland Southeast Asia and north-east India.

National Conservation Status Least Concern; Widespread but Uncommon.

IUCN Red List Status Least Concern.

Larva To the authors' best knowledge, undescribed; should be typical of genus.

Marcus Ng

Dorsal view of a male, showing the iridescent wing markings.

Marcus Ng

Male in typical pose, with the wings slightly depressed and at an angle.

Cheong Loong Fah

Female with hyaline patches on the wings.

Robin Ngiam

Pair in wheel.

YELLOW-BARRED FLUTTERER *Rhyothemis phyllis* (Sulzer, 1776)

Size HWL: 33–37mm; TBL: 39–41mm

Description Moderately large, highly aerial dragonfly with distinctive colours. Eyes reddish-brown and grey-brown. Thorax and abdomen metallic bronzey-green. Wings very long, with black on nodus and tips. Hindwing-base greatly expanded, with alternating dark brown (with slight blue-violet reflex) and yellow patches, which may serve to mimic a hymenopteran in flight. Female similar to male but abdomen stouter.

Habitat & Habits Occurs at marshy and well-vegetated ponds, lakes, reservoirs, grassland and drains in semi-urban to urban open country and near coasts. Both sexes often gather in feeding swarms, sometimes with other species, on hilltops or in forest clearings and edges. Most often encountered foraging in the air, but flutters down to settle on low vegetation during cloudy weather, permitting close approach.

Presence in Singapore The most widespread and abundant of its genus locally, recorded island-wide in nature reserves and nature parks as well as urban parks, gardens and fields in urban areas.

Etymology Phyllis was a Thracian princess who married the king of Athens but hanged herself when her husband failed to return to her as promised.

Distribution Southeast Asia and Australasia, reaching Guam and Fiji. Also Taiwan. Several regional subspecies in need of taxonomic review.

National Conservation Status Least Concern; Widespread and Common.

IUCN Red List Status Least Concern.

Larva Typical libellulid in appearance. Squat with very long hindlegs, femur reaching abdominal segment 7.

Marcus Ng

Female perched on a twig during cloudy weather.

Dorsal view of a male, showing the wing markings.

Dorsal view of a female, with its relatively shorter abdomen.

Lateral view of a male, showing the metallic thorax.

Sapphire Flutterer *Rhyothemis triangularis* Kirby, 1889

Size HWL: 21–24mm; TBL: 23–26mm

Description Smallish dragonfly with stubby abdomen and unmistakable wing colours. Eyes dark brownish-red and brown-black. Male has dark steely-blue thorax and abdomen. Hindwing-base greatly expanded, with rear margin reaching abdominal segment 6. Basal quarters of both wings bear sapphire-blue patches with irregular outer margins. Female similar but darker – more blackish than blue – with comparatively shorter abdomen.

Habitat & Habits Found around well-vegetated drains, ponds and sluggish streams in swampy forests, and marshy edges of reservoirs and urban parks. Flight can be fluttery or zippy. Males actively give chase to each other and other dragonflies on sunny days. Fond of perching with wings facing the sun. Like other *Rhyothemis* species, tends to 'wave' its wings gently right after landing on perch before settling down fully.

Presence in Singapore Recorded in many locations, including the Bukit Timah, Central Catchment and Labrador Nature Reserves and adjacent nature parks such as Windsor Nature Park, Kent Ridge Park, Toa Payoh Town Park, Tampines Eco-Green, the Singapore Botanic Gardens, the Istana and Pulau Ubin.

Distribution Sundaland and mainland tropical Asia.

National Conservation Status Least Concern; Widespread but Uncommon.

IUCN Red List Status Least Concern.

Larva Typical of genus.

Marcus Ng

Male basking at the pond near the entrance to Windsor Nature Park.

Marcus Ng

The broad hindwings give the impression of a stubby insect.

Tang Hung Bun

A female.

Marcus Ng

A very dark and worn female photographed in Malaysia.

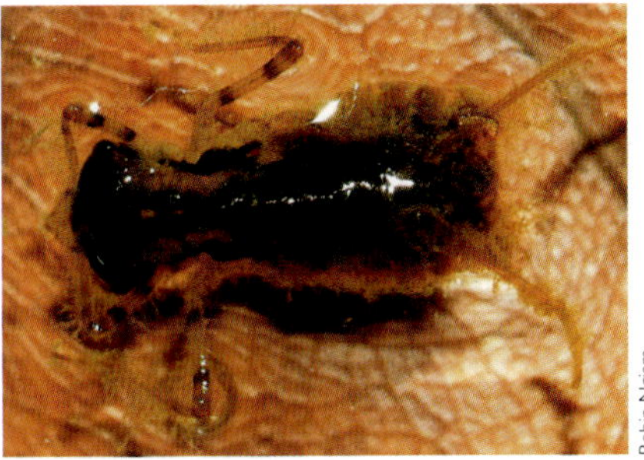

Robin Ngiam

The squat larva.

POTBELLIED ELF *Risiophlebia dohrni* (Krüger, 1902)

Size HWL: 18–19mm; TBL: 24–25mm

Description Very small dragonfly with unique eye patterns and slender, arched abdomen. Eye dark reddish-brown and yellowish-white, with vertical brown streak. Thorax dark metallic green with yellow stripes that extend to abdominal segment 3. Rest of abdomen dark with slight yellow banding and dorsal markings; segments 1–3 greatly inflated; remaining segments slender and curved in profile. Wings hyaline and narrow, with no trace of anal loop. Female similar to male.

Habitat & Habits Found in dense swampy forests. Males guard small pools under a dense canopy, illuminated by narrow shafts of sunlight. Females oviposit in these sites, guarded by males. Perches very low and inconspicuously on emergent vegetation. May occasionally bask at edges of swampy forests.

Presence in Singapore Recorded in Nee Soon Swamp Forest. Also sighted at Thomson Nature Park in 2016.

Etymology Originally placed in *Nannophlebia*, a genus now restricted to eastern Indonesia and Australasia. John Cowley then reassigned it to *Risiophlebia* in 1934, combining the name of Friedrich Ris (a Swiss doctor who had redescribed the dragonfly as *Oda dohrni* in 1909) with *phlebia* (from *phleps*, Greek for 'veins'), a common suffix in odonate names. Specific epithet honours Dr Heinrich Wolfgang Ludwig Dohrn, a German entomologist and politician.

Distribution Sundaland. Reports of the species from Vietnam are now believed to be of *R. guentheri* (Kosterin, 2015), a species restricted to southern Indochina.

National Conservation Status Endangered; Restricted and Rare.

IUCN Red List Status Least Concern.

Larva Unknown for genus.

Marcus Ng

Lateral view of a male, showing the unique eye markings and arched abdomen with a greatly expanded base.

Marcus Ng

Female at the edge of swampy forest.

Tang Hung Bun

Female with its slightly thicker abdomen.

Fiora Li

Pair in copula, photographed using a smartphone.

Marcus Ng

Mating pair near a swampy forest stream.

ELF *Tetrathemis hyalina* Kirby, 1889

Size HWL: 20–22mm; TBL: 24–26mm

Description Small dragonfly with brilliant blue-green eyes. Wings narrow and hyaline, but may have slight yellow basal tint. Male's thorax dark metallic green with broad yellow stripes. Abdomen mostly dark, with yellow lateral markings on segments 1–5 and paired yellow dorsal spots on segment 7. Appendages black. Base of abdomen not expanded, unlike in the similarly sized Potbellied Elf (p. 302). Female similar but with thicker abdomen and more extensive yellow tint on wings. Previously listed as subspecies *T. irregularis hyalina* (Kirby, 1889) – an error made by Ris in 1909 that was unfortunately perpetuated by subsequent odonatologists. Now regarded as distinct species from *T. irregularis*. The two species differ in shape of tip of upper abdominal appendage.

Habitat Found at small pools and slow streams in shaded forests and swamp forests. Very localized and inconspicuous on account of its small size and habit of perching in fairly dark corners. Males guard small forest pools for hours, mating with females that arrive through the day. Females oviposit by adhering eggs to tips of small branches, roots or even spiders' webs overhanging water. Also basks and forages in the canopy, which adds to the difficulty of finding it.

Presence in Singapore Restricted to the Central Catchment Nature Reserve.

Etymology Generic epithet combines the Greek for 'four' (*tetra*) with *themis*, a common appellation in skimmer names. It refers to an important cell in the forewing that is triangular in most other true dragonflies but four-sided in this genus. Specific epithet refers to the somewhat hyaline wings.

Distribution Sundaland, the Philippines and mainland Southeast Asia. Taxonomic review of genus is required.

National Conservation Status Endangered; Restricted and Very Rare.

IUCN Red List Status Least Concern.

Larva Typical libellulid in appearance. Small (*c.* 10 mm) and squat. Found in bottom detritus.

Erland Refling Nielsen

Male photographed in Taman Negara, Malaysia.

Rory Dow

Male anal appendages of Tetrathemis irregularis.

Rory Dow

Male anal appendages of Tetrathemis hyalina.

Cheong Loong Fah

Female with a fairly extensive tint at the wing-bases.

Cheong Loong Fah

Female with less colour in the wings.

Marcus Ng

Female in Golf Link.

Cheong Yi Wei

Female ovipositing on a strand of spider's web.

Keith Wilson

Female ovipositing on a twig. Photo taken in Endau-Rompin, Malaysia.

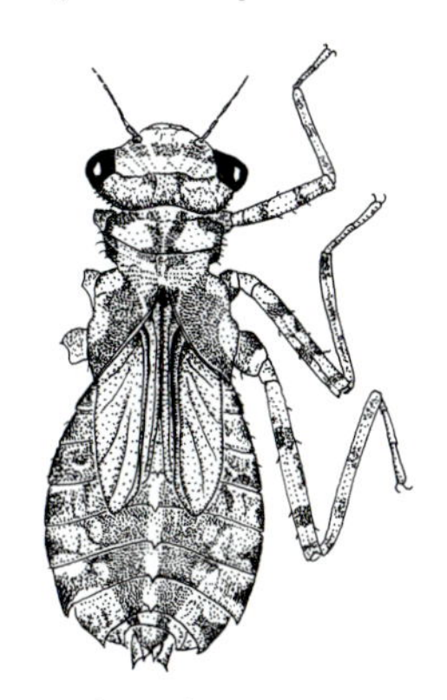

Albert G. Orr.

Drawing of a Tetrathemis larva.

White-barred Duskhawk *Tholymis tillarga* (Fabricius, 1798)

Size HWL: 34–35mm; TBL: 44–47mm

Description Medium-sized reddish dragonfly with distinctive wing markings and crepuscular habits. Eyes reddish-brown and white-green; paler in female. Male's thorax and abdomen dull red. Hindwing-base moderately expanded; 'toe-tip' of anal loop meets rear wing-margin. Hindwing features amber-brown suffusion that ends at nodus, beyond which is a small gleaming white patch. Female yellowish-brown; wings hyaline with light brown tint before nodus. Immature male similar to female.

Habitat & Habits Found at ponds, lakes, canals and drains with still or slow-flowing water. Occurs at forest edges as well as in disturbed and urban habitats. Crepuscular, hiding amid dense vegetation in hanging position by day. If flushed, makes short, erratic flight before settling at another spot. Becomes active in the late afternoon, from about 5 p.m., when males appear at waterbodies, patrolling their territories in a regular circuit and stopping often to hover. Rivals that encounter each other engage in furious aerial pursuits before returning to their beats. The males remain until last light, when their pale wing-patches are all that traces their path over the water. Mating and oviposition take place shortly before sunset. Female oviposits on shallow aquatic vegetation while male hovers nearby, guarding. After dipping her abdomen into the water to release some eggs, female makes rapid 180-degree turns before releasing another batch. This characteristic movement gave rise to the species' other common name, Twister. With total darkness, the dragonflies seek out roosts in dense vegetation, hanging as they do by day. They may be attracted to building lights, but unlike other dragonflies are apparently not attacked by geckos, as their wing colours are thought to suggest a stinging hymenopteran insect.

Presence in Singapore Recorded across the island, even at large urban canals, where males can be seen flitting about towards dusk.

Etymology That of genus unclear, but suggested to be a compound of ***tho**rax*, ***ly**gaios* (Greek for 'shadowed') and *the**mis*** to form *Tholymis*. Hagen, who coined the genus in 1867 for *T. citrina*, may have been referring to the darkish thorax of this Neotropical species. Etymology of *tillarga* a mystery.

Distribution Old World tropics from Africa to Australasia and Micronesia.

Marcus Ng

Dorsal view of a male, with strands of spiders' webs stuck to the wings.

National Conservation Status Least Concern; Widespread and Common.

IUCN Red List Status Least Concern.

Larva Fairly slender; abdomen lacks long spines but has dark dorsal markings; appendages longish. Dwells at bottoms of ponds and slow-flowing waterbodies.

Marcus Ng

Dorsal view of a young male with pale colours. Note the open-toed anal loop.

Marcus Ng

Dorsal view of a female.

Erland Refling Nielsen

The white wing-patches are especially prominent during patrol flights. Photo taken in Malaysia.

Robin Ngiam

The larva showing the longish anal appendages.

SADDLEBAG GLIDER *Tramea transmarina euryale* Selys, 1878

Size HWL: 42–45mm; TBL: 53–54mm

Description Large, reddish and highly aerial dragonfly. Eyes dark brown and grey-black. Male's thorax dark brown and hairy. Abdomen dull red with black dorsal markings on segments 8–9. Appendages black and very long, especially upper pair. Wings long and hyaline, with reddish basal veins. Hindwing-base well expanded, with small dark brown basal patch. Female similar but abdomen browner. May be confused with the Scarlet Basker (p. 319), but that species is smaller, brighter red, and has different eye colours and shorter appendages. Distinguished in the air from the Wandering Glider (p. 282) by darker colours and brown hindwing-patch.

Habitat & Habits Occurs around well-vegetated ponds, lakes and drains in open country, near forests or urban parks; also landward sides of mangroves. Adults sail over waterbodies, forest clearings and hilltops, though never in large swarms, sometimes in the company of Wandering Gliders. Seldom perches, but may land on tips of reeds and twigs during overcast weather, usually on a high spot. Habit of slanting abdomen downwards when soaring or perched. Migratory species that wanders far from breeding sites. Species in this genus practise an unusual method of oviposition: after mating, the pair remains in tandem as they fly, with female disengaging at times to dip its abdomen and eggs into the water before flying back to the male, which then grasps her again. The pair then continues to fly in tandem to find another oviposition site.

Presence in Singapore Recorded in various locations such as MacRitchie Reservoir, Sungei Buloh Wetland Reserve, Labrador Nature Reserve, Sengkang Riverside Park, Admiralty Park, Jurong Lake Gardens, the Singapore Botanic Gardens, Pasir Ris Park, Pulau Ubin and Pulau Semakau.

Etymology Genus originally proposed by Hermann August Hagen as *Trapezostigma*, referring to trapezoid-shaped pterostigma, but was eventually published in 1861 in a shortened form, *Tramea*, which is a pun on

Marcus Ng

Lateral view of a male, showing the long legs and long anal appendages.

trameare, 'to pass through' in Latin – an apt name, given the migratory habits of this genus. Specific epithet means 'across the sea'. Euryale was one of three gorgons (sisters with snakes for hair) of Greek mythology.

Distribution South China, Taiwan, Japan, Southeast Asia and Australasia, reaching Polynesia. Several subspecies in need of taxonomic review.

National Conservation Status Least Concern; Widespread and Common.

IUCN Red List Status Least Concern.

Larva Similar to larva of the Water Monarch (p. 245), but lacks dorsal spines.

Marcus Ng

Male, showing the head colours and abdominal markings.

Henrietta Woo

Male in flight, showing the coloured patch on the hindwing-bases.

Crimson Dropwing *Trithemis aurora* (Burmeister, 1839)

Size HWL: 25–27mm; TBL: 32–35mm

Description Small-medium dragonfly with fiery colours and conspicuous habits. Male has red eyes that are darker below. Body and wing veins glowing pinkish-red; hindwings have fairly extensive light brown basal tint. Female has brown and grey-white eyes. Body light yellow-brown with dark, mostly incomplete stripes on sides of thorax and thin dark streaks along abdomen, which become thicker towards tip. Wings hyaline but with extensive light brown tint at hindwing-base; compare with female White-tipped Demon (p. 247), which has reduced colour at wing-base. Abdominal markings differ from female Restless Demon's (p. 249). Immature male similar to female but without dark abdominal markings.

Habitat & Habits Occurs at vegetated ponds, lakes, drains and slow-flowing streams in open country and near forests. Tolerates disturbed habitats. Both sexes perch on emergent vegetation by or near water, usually with wings depressed in typical dropwing manner, even when in obelisk position. Also basks at forest edges and clearings, especially immatures. Active at breeding sites from late morning until early evening. Seen to roost vertically on vegetation at night.

Presence in Singapore Occurs island-wide in nature reserves, nature parks, urban parks and gardens.

Etymology Generic epithet combines *tri*, referring to three lobes on rear margin of prothorax, and *themis*, a common suffix in libellulid genera. Aurora is the Roman goddess of the dawn, who fills the sky at daybreak with colours that are reflected in this gorgeous dragonfly.

Distribution Widespread in tropical and subtropical Asia.

National Conservation Status Least Concern; Widespread and Common.

IUCN Red List Status Least Concern.

Larva Typical libellulid in appearance. Pentagon-shaped head; abdomen ovate. Inhabits bottom detritus.

Marcus Ng

Dorsal view of a mature male in obelisk.

Marcus Ng

Lateral view of a female. Note the thoracic markings and extensive tint on the hindwing-bases (compare with the female Indigo Dropwing and Indothemis *species).*

Marcus Ng

Young male, with the thoracic markings still visible, which is beginning to turn pink.

Marcus Ng

Lateral view of a mature male, showing the fairly extensive hindwing-patch.

Marcus Ng

Dorsolateral view of the female.

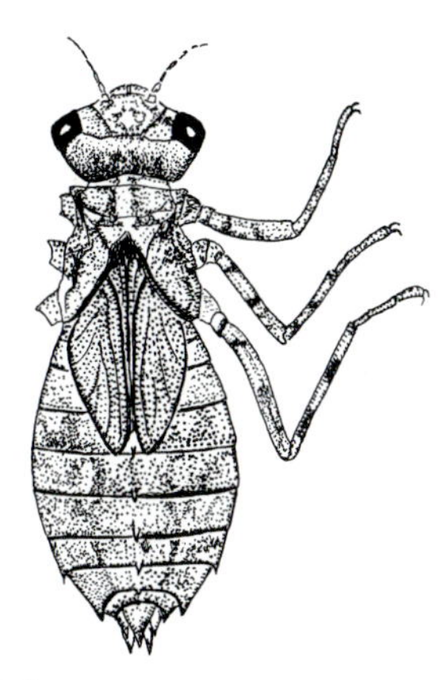

Albert G. Orr

Drawing of the larva.

INDIGO DROPWING *Trithemis festiva* (Rambur, 1842)

Size HWL: 26–28mm; TBL: 33–37mm

Description Small-medium, dark blue dragonfly associated with flowing water. Male's eyes dark brown and grey-black, with bluish rear edging. Thorax and dorsum of basal abdominal segments (up to segment 4) indigo blue. Rest of abdomen dark, with yellow dorsal streaks that may be faint. Wings hyaline, with small, irregular brown patch at hindwing-base; compare with rounded border of hindwing-patch in the Restless Demon (p. 249). Female has brown and whitish-grey eyes, and yellow-brown body with thick dark markings on sides of thorax and along abdomen. Wings hyaline, with small light brown tint on hindwing-base. Distinguished from the female Crimson Dropwing (p. 310) by thicker thoracic markings and much smaller hindwing-patch. Distinguished from the female Restless Demon by thicker, better defined dark thoracic markings as well as habitat – this species inhabits streams, while the Restless Demon prefers still waterbodies.

Habitat & Habits Occurs at streams, rivers and drains with fast, clear water in open country, forest edges or urban habitats. Males perch on rocks, sand bars or streamside vegetation around midday, chasing rivals and seizing arriving females. Copulation is brief, and immediately after female oviposits by flicking her abdomen at the water. Males may also forage further afield, in clearings or along trails, later in the day or during cloudy weather. Females less often seen, staying in the canopy or vegetation a little away from water until they are ready to mate.

Presence in Singapore Recorded in various locations, including the Central Catchment and Bukit Timah Nature Reserves, Bukit Batok Nature Park, Bishan-Ang Mo Kio Park, HortPark, Istana, Telok Blangah Hill Park, the Rail Corridor, the Western Catchment, Bukit Brown and Pulau Ubin.

Distribution Old World tropics from the eastern Mediterranean to Southeast Asia and South China.

National Conservation Status Least Concern; Widespread and Common.

IUCN Red List Status Least Concern.

Larva Typical of genus. Lighter coloured compared to the Crimson Dropwing.

Marcus Ng

Lateral view of a male.

Marcus Ng

Dorsal view of a female. Compare with the female Restless Demon, which has more colour in the wing-bases and a thinner dorsal stripe on the thorax.

Marcus Ng

Lateral view of a female. Note the well-defined thoracic stripes (compare with the female Crimson Dropwing and White-tipped Demon).

Tang Hung Bun

Pair in wheel. Copula lasts for only a minute or so.

Tang Hung Bun

The relatively pale larva.

Dancing Dropwing *Trithemis pallidinervis* (Kirby, 1889)

Size HWL: 30–33mm; TBL: 41–44mm

Description Pale, long-legged dragonfly associated with grassy habitats. Eyes light brown and greyish-white; darker in males than females. Male has shiny violet frons. Thorax pale yellow, with four dark lines – two complete, two nearly so – on each side. Older males develop brown-grey bloom on thorax. Abdomen black with yellow streaks up to segment 7. Segments 8–10 mostly black. Appendages yellow, tipped with black. Wings hyaline but with pale yellow veins that seem to glisten in sunlight. Hindwing-base has small amber-brown patch. Female similar to male but frons yellow. Easily distinguished from other dropwings by longer legs, wing colours and perching position, which is never with depressed wings.

Habitat & Habits Found at reservoirs, lakes, marshes and open, grassy areas, especially near the coast, but also in urban areas. Seldom at small ponds. Both sexes typically perch at tip of a twig or reed, swaying in the breeze like ballet dancers on tiptoe, hence the common name. Wings never held depressed, unlike those of other dropwings; usually kept wide open but may be clasped over abdomen, like a damselfly's wings, during strong headwinds.

Presence in Singapore Occurs island-wide in open, grassy areas around larger reservoirs, coastal parks and large, open wetlands, and at Pulau Ubin.

Etymology Specific epithet, derived from *pallidus* (Latin for 'pale') and *nervis*, refers to pale wing veins (nerves).

Distribution Tropical mainland Asia and much of island Southeast Asia.

National Conservation Status Least Concern; Widespread but Uncommon.

IUCN Red List Status Least Concern.

Larva Typical appearance of genus. Found among aquatic vegetation.

Marcus Ng

Male clinging to the tip of a reed. The wings are 'closed' as the dragonfly typically perches against a headwind.

Marcus Ng

Male with a darkening thorax and showing the violet frons.

Marcus Ng

A female. Note the pale frons and the pale primary wing veins.

Tang Hung Bun

Pair in wheel.

Marcus Ng

Head-on view of a female clinging to a grass blade.

TREEHUGGER *Tyriobapta torrida* Kirby, 1889

Size HWL: 25–26mm; TBL: 29–32mm

Description Small-medium blue dragonfly with unique patterns and 'treehugging' habit. Male has dark brown and blackish eyes, with paler edging. Thorax and abdomen dull blue. Hindwing distinguished by large dark and iridescent basal patch with green-purple reflex. Female has brown and pale grey-brown eyes. Body light brown with strong dark banding on abdomen. Wings hyaline but with slightly darkened tips. Young male similar to female, with just a hint of colour at hindwing-base.

Habitat & Habits Found at slow-flowing forest streams and swampy forests. Both sexes perch, often in small groups, on tree trunks or large rocks along forest trails and clearings, at times far from water. Strong site fidelity, with individuals coming back to perch on the same tree over a long period. May also use fence posts, passing human legs and other vertical surfaces as perches. Breeds in small, shaded forest pools or quiet side pools of small streams. At breeding sites, male occupies a twig above the water. Rivals are greeted with a non-contact agonistic display, in which males face off in mid-air, darting at each other and displaying their hindwing-patches, which glisten in the sunlight as they rise in the air. After copulation, female oviposits by repeatedly flicking her abdomen at the water's surface to propel her eggs along with water droplets on to a nearby bank. Male guards female, tracing near rectangular flight path above her and hovering briefly at the 'corners'.

Presence in Singapore Recorded in the Central Catchment and Bukit Timah Nature Reserves, adjacent nature parks, Bukit Brown and the Western Catchment.

Etymology Generic epithet combines the Greek terms for a purple cloth from the city of Tyre (*tyrius*) and 'to dye' (*bapto*). Ancient Tyre was famous for its purple textiles, which were made using a dye from *Murex* snails. Kirby may have compared the wing colours with this cloth. Specific epithet may refer to dragonfly's origins in the torrid or equatorial regions, which ancient Greeks such as Aristotle believed to be uninhabitable to people.

Distribution Sundaland.

Marcus Ng

Mature male on a tree trunk by a forest trail.

National Conservation Status Least Concern; Widespread and Common.

IUCN Red List Status Least Concern.

Larva Known but not formally described. Ovate abdomen with long hindlegs. Distinctive dark patterns on labial palps. Found among bottom leaf litter.

Marcus Ng

Young male with undeveloped wing colours.

Marcus Ng

Very young male with patterns that resemble the female's.

Marcus Ng

Female 'hugging' the wall of a forest hut.

Marcus Ng

Lateral view of a female, showing the 'tiger' stripes and thickish abdomen.

Robin Ngiam

The darkish larva.

Rare Basker *Urothemis abbotti* Laidlaw, 1927

Size HWL: 36–38mm; TBL: 42–46mm

Description Moderately large red dragonfly, very similar to the Scarlet Basker (opposite). Male's eyes dark brown. Thorax yellowish-brown (based on holotype, which is probably an immature individual); abdomen red with black dorsal flecks. Appendages red. Wing-base has large brownish patch with yellow border. Female brownish-yellow; wing-base patches less extensive. Distinguished from Scarlet by larger size and more extensive wing-base patch, especially on forewing, where markings may reach second antenodal crossvein and cover entire median sector cell.

Habitat & Habits Very little known. Thought to be a mangrove species but most probably also occurs in open habitats such as ponds, lakes and slow streams. Behaviour probably similar to Scarlet's.

Presence in Singapore Known locally from single incomplete specimen mentioned in Laidlaw when he described the species in 1927 and surmised 'that it is an inhabitant of the mangrove swamps'. Apparently it was collected by J. C. Moulton, former director of the Raffles Museum. Lieftinck (1954) reported the species to be 'fairly common around Singapore', but there are no specimens to substantiate this claim. A futile search for validated museum specimens resulted in the removal of this species from the Singapore checklist in 2016. However, more recently, the specimen mentioned by Laidlaw was digitized and made publicly available by the Natural History Museum, London. Hence it has been reinstated. Despite Lieftinck's claim that this mysterious species was common and widespread in Singapore and Malaya, there is a paucity of records for it. It may be overlooked due to its similarity with the very common Scarlet.

Etymology Named after Dr William Louis Abbott (1860–1936), an American physician who studied and collected Southeast Asian wildlife in 1899–1909, using a Singapore-based ship named *Terrapin*. Abbott collected the holotype in Thailand.

Distribution Thailand (Trang), Vietnam (Hanoi), Peninsular Malaysia and Singapore (formerly).

National Conservation Status Extinct.

IUCN Red List Status Vulnerable.

Larva Unknown, but should be similar to that of Scarlet.

Specimen collected in Singapore in 1921. Reproduced under a Creative Commons License from the Natural History Museum, London. www.gbif.org/occurrence/1826387004.

SCARLET BASKER *Urothemis signata insignata* (Selys, 1872)

Size HWL: 34–35mm; TBL: 42–45mm

Description Medium-sized red dragonfly with erect bearing and long, pointed wings with open venation. Male's eyes red and blackish. Thorax, abdomen and appendages bright red, with two small black spots on dorsum of segments 8–9. Veins along leading edges of wings red up to nodus. Hindwing-base has prominent dark brown patch with paler outer margin. Distinguished from the closely related Coastal Glider (p. 255) by larger size, bicolored hindwing-patch and abdominal markings. Distinguished from other medium-sized red dragonflies (*Crocothemis*, *Rhodothemis* and *Orthetrum* species) by erect posture, pointed wings with open venation, and abdominal markings. Female has reddish-brown and greyish-white eyes. Thorax and abdomen light yellowish-brown with row of dark dorsal markings on abdomen. Hindwing-base has bicoloured patch similar to male's. Andromorphs not uncommon. Distinguished from female Coastal Glider by dorsal markings of abdomen (more clearly defined in Coastal Glider) and long pseudo-ovipositor below abdominal segment 8.

Habitat & Habits Occurs at ponds, marshes and reservoirs in open country, forest edges and urban areas, and near the coast. Sun-loving species that perches conspicuously, often in obelisk position, on tips of projecting twigs and reeds, and actively chases off other dragonflies, especially red ones. Females usually forage further from water than males, on tops of trees, bushes or masses of fallen branches.

Presence in Singapore Recorded island-wide. Known sites include the Central Catchment and Bukit Timah Nature Reserves, Sungei Buloh Wetland Reserve, Kranji Marshes, Bishan-Ang Mo Kio Park, Pasir Ris Park, HortPark, Tampines Eco-Green, the Singapore Botanic Gardens and Pulau Ubin.

Etymology Generic prefix probably derived from *oura* ('tail' in Greek), referring to female's elongated vulvar scale (pseudo-ovipositor), which extends from abdominal segment 8 to 10, and serves as a chute for the eggs. Specific epithet means 'marked with' and may refer to black dorsal spots on abdomen or hindwing basal patch.

Marcus Ng

Lateral view of a male, showing the eye colours, abdominal markings and typical perching pose.

Distribution Subspecies *insignata* restricted to Sundaland. Various other subspecies (needing taxonomic review) range from tropical Asia to Australasia.

National Conservation Status Least Concern; Widespread and Common.

IUCN Red List Status Least Concern.

Larva Triangular head with backwards pointing eyes. Abdomen ovate and light brown; legs banded.

Marcus Ng

Lateral view of the female, showing the long vulvar scale extending from abdominal segment 8.

Marcus Ng

Dorsal view of the male, showing the wing-base colours.

Marcus Ng

Dorsal view of the female, showing the dark abdominal markings.

Marcus Ng

Andromorph female.

Robin Ngiam

The light brown larva.

WHITE DUSKDARTER *Zyxomma obtusum* Albarda, 1881

Size HWL: c. 38mm; TBL: c. 42mm

Description Medium-sized, wraith-like dragonfly. Male's eyes greyish-white; body entirely covered by white pruinescence. White bloom extends into wing veins; wing-tips black. Abdominal segments 1–3 slightly swollen, appendages dark. Female has light green eyes, light brown thorax and brownish abdomen with darker banding. Base of abdomen slightly swollen. Wings hyaline with slight brownish-amber tint at hindwing-bases and slightly darkened wing-tips. Young male similar to female.

Habitat & Habits Occurs at drains, ponds and lakes in open country. Crepuscular, perching in dense vegetation by day, and becoming active in the early evening and pre-dawn hours, flying fast and low above waterbodies. Males conspicuous as they flit about and give chase in near darkness. Attracted to lights at night. Females seen to oviposit on floating branches, forming white egg mass.

Presence in Singapore To date, recorded only in Pulau Ubin but possibly overlooked on main island.

Etymology Generic epithet combines *zeuxis* ('yoking' in Greek) and *omma* ('eye'), referring to large, contiguous eyes of genus, which resemble crepuscular hawkers (Aeshnidae) in both their appearance and habits. Specific epithet means 'dull' in colour in Latin.

Distribution Southeast Asia, southern China, Taiwan and Japan.

National Conservation Status Critically Endangered; Restricted and Uncommon.

IUCN Red List Status Least Concern.

Larva Pentagon-shaped head with small, backwards pointing eyes. Abdomen elongated.

Erland Refling Nielsen

Patrolling male showing its pale colours. Photo taken in North Sulawesi.

Marcus Ng

Pair (female on the right) at a forest-edge pond in Maliau Basin, Malaysia.

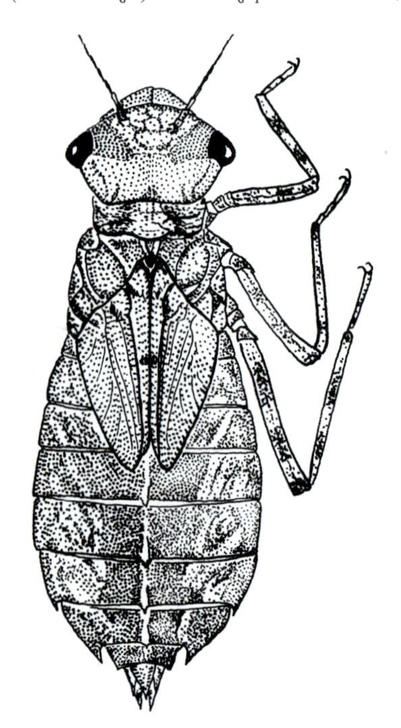
Albert G. Orr

Drawing of the larva.

Marcus Ng

The male in flight presents a ghostly visage.

Marcus Ng

The darker female is usually detected while being pursued by the male.

SLENDER DUSKDARTER *Zyxomma petiolatum* Rambur, 1842

Size HWL: 31–33mm; TBL: 49–52mm

Description Medium-sized, very slender dragonfly that may be mistaken for a small hawker. Male has apple-green eyes. Body dark brown to dark grey; abdomen very slender, but swollen at base. Female similar, but abdomen slightly thicker. Wings hyaline with slight dark tinge at bases. Wings become darkened with age in both sexes.

Habitat & Habits Found at weedy drains, ponds and streams in open country and forest edges. Perches in trees and dense vegetation by day, often with abdomen hanging downwards, like an aeshnid. Becomes active towards sunset until long after dusk. Males fly rapidly around their territory, hovering briefly at certain regular spots before continuing their patrol. Attracted to lights at night.

Presence in Singapore Occurs across the island in both forests and urban areas, but probably overlooked due to its crepuscular habits.

Etymology Specific epithet comes from *petiolus*, Greek for a 'petiole' or 'leaf-stalk', and refers to exceedingly slender abdomen.

Distribution Tropical Asia, north to Japan, and Australasia.

National Conservation Status Least Concern; Widespread and Common.

IUCN Red List Status Least Concern.

Larva Similar in appearance to that of the White Duskdarter (p. 321), but with two distinctive pale spots on abdominal segment 9.

Marcus Ng

Male perched on a tree branch at midday.

Marcus Ng

Patrolling male at sunset. Photo taken in Vietnam.

Marcus Ng

Female that was drawn to lights at a housing flat.

Marcus Ng

Female with darkened wings.

Robin Ngiam

The larva, showing pale spots on abdominal segment 9.

Macromiidae (Cruisers)

This family consists of medium-sized to very large dragonflies with bright green or bluish eyes that meet broadly. The thorax is fairly robust, usually dark and metallic, and bears fairly long legs. The wings are largely hyaline, very long and pointed, and feature an anal loop that is never sock shaped as in the Libellulidae or Corduliidae. Males have an acute hindwing anal angle. They perch in a hanging position, similarly to hawkers.

Along with the day-active hawkers (*Anax* species) and broad-winged skimmers, such as *Camacinia*, *Pantala* and *Tramea* species, cruisers are among the most aerial of local dragonflies. They are fast and commanding in the air, hunting and patrolling their territories for long periods, and resting only when it becomes overcast. Adults are usually found in forests and swampy areas, cruising conspicuously over streams and trails, or darting about over the canopy.

Cruisers were formerly placed in the family Corduliidae, but are now regarded as a separate family. Their larvae are similar to those of skimmers (Libellulidae) but are rather flattened and have very long legs that give them a 'spidery' appearance. There are at least 123 species worldwide, and three in Singapore.

The name of the family and its largest genus, *Macromia*, is derived from the Greek terms *makros* ('long') and *omos* ('shoulder'). It refers to the costa or leading edge of the forewings, in which the humeral ('shoulder') section before the nodus is much longer than the cubital or 'forearm' edge after the nodus.

Marcus Ng

Head and wings of a male Pond Cruiser, showing the large green eyes and very long 'shoulder' (the leading edge of the forewing before the nodus) that distinguishes the family.

Pond Cruiser *Epophthalmia vittigera* (Rambur, 1842)

Size HWL: 50–52mm; TBL: 78–82mm

Description Very large, dark dragonfly with brilliant blue-green eyes. Wings long, pointed and hyaline, except for slight dark tint at leading edges of wing-bases. Body dark purplish and metallic with yellow markings on thorax and basal abdominal segments. Male appendages black and pronounced. Female similar to male, but with rounded hindwing-base.

Habitat & Habits Occurs at lakes, marshes and swamps in semi-open country, forest edges and mangroves. Males command very large territories, and are conspicuous as they patrol the edges of lakes and reservoirs, or cruise backwards and forwards along trails, often gliding and little bothered by passers-by. They seek out masses of fallen branches during overcast weather and perch in a hanging position, at times in small groups.

Presence in Singapore Recorded in various locations, such as the Bukit Timah, Central Catchment and Labrador Nature Reserves, the nature parks, Sungei Buloh Wetland Reserve, Kranji Marshes, Kent Ridge, Singapore Botanic Gardens and Pulau Ubin.

Etymology Generic epithet combines the Greek for 'on' (*epi*) and 'eye' (*ophthalmos*), and refers to small process on rear margin of each eye. Specific epithet thought to combine *vitta* (a head-band worn by Roman priests) and *gero* ('to wear'), and refer to the thoracic markings.

Distribution India to Southeast Asia.

National Conservation Status Least Concern; Widespread and Common.

IUCN Red List Status Least Concern.

Larva Large and robust, with long legs. Head rectangular, with small eyes. Slow-moving bottom dweller among leaf litter.

Marcus Ng

Dorsal view of a male, showing the yellow markings on the thorax and abdomen. Note also the rather acute base of the hindwing.

Marcus Ng

Pond Cruisers may patrol trails for hours, settling only during overcast periods.

Marcus Ng

Lateral view of a male in a typical perching position.

Marcus Ng

Male at Sungei Buloh Wetland Reserve.

Robin Ngiam

The rather spidery larva.

STREAM CRUISER *Macromia cincta* Rambur, 1842

Size HWL: 43–47mm; TBL: 62–69mm

Description Large dark dragonfly with bright blue-green eyes. Male has dark, metallic body and strong white stripe on sides of synthorax, continuing into dorsum and forming prominent white 'saddle', distinguishing this species from the Pond Cruiser (p. 326). Abdomen has white band on segment 2, and faint banding thereafter. Abdomen-tip swollen and club-like, with prominent appendages. Female similar, but darker and with stronger dark tint at wing-bases. Hindwing-base rounded.

Habitat & Habits Found at slow-flowing streams in swampy forests. Males patrol streams and shaded portions of trails, and may also dart about over the canopy at high speeds, chasing rivals. During overcast weather both sexes may descend to seek out perches on low vegetation. Females may hunt along shaded trails near swamps. They oviposit alone in shaded streams in swampy forests, hovering while repeatedly stabbing the abdomen into the water to release batches of eggs.

Presence in Singapore Recorded in the Central Catchment Nature Reserve, Windsor and Dairy Farm Nature Parks.

Etymology Specific epithet derived from *cinctus*, Latin for 'bordered' or 'surrounded'. It may refer to white band that runs from side of thorax and across dorsum.

Distribution Sundaland and Cambodia.

National Conservation Status Endangered; Restricted and Rare.

IUCN Red List Status Least Concern.

Larva Robust and long legged, with long setae on hindleg tibia. Benthic and found amid decaying vegetation.

Marcus Ng

A male. Note the lack of yellow markings, which distinguish this species from the Pond Cruiser.

Marcus Ng

Female has a darker body than the male and dark streaks at the bases of the wings.

Tang Hung Bun

Lateral view of a male, showing the prominent white stripe and 'saddle' on the thorax.

Marcel Finlay

Male cruising down a forest trail. During such patrol flights, the male may break off to pursue a rival high above the forest canopy.

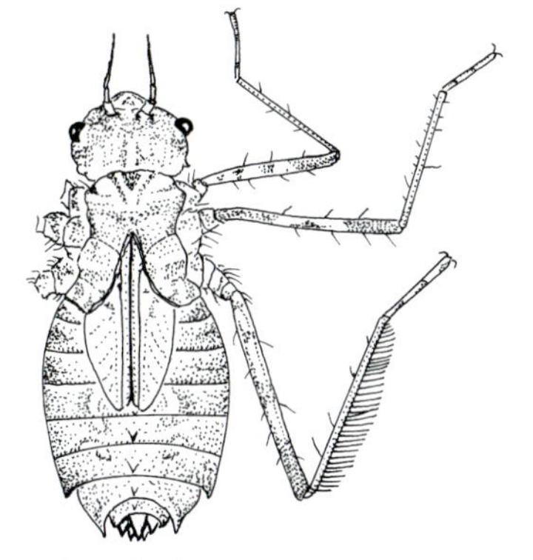

Albert G. Orr

Drawing of the spidery larva.

LESSER STREAM CRUISER *Macromia cydippe*
Laidlaw, 1922

Size HWL: 39–41mm; TBL: 58–62mm

Description Fairly large, dark dragonfly with bright green eyes. Body dark metallic green, with yellow stripe on sides of thorax and distinct yellow band on abdominal segment 7. Female similar to male but has rounded hindwing-base. Distinguished from the Stream Cruiser (p. 328) by smaller size, body markings and male appendages (lower appendage longer than uppers).

Habitat & Habits Found at sluggish, clear streams in forests and swampy forests.

Presence in Singapore Rare and highly fugitive species. Recorded in the Central Catchment Nature Reserve and Windsor Nature Park.

Etymology Cydippe is the name of a nereid (sea-nymph) in Greek mythology.

Distribution Sundaland and central Thailand.

National Conservation Status Endangered; Restricted and Rare.

IUCN Red List Status Least Concern.

Larva Legs shorter than those of other *Macromia* species, and folded towards body to reduce surface exposure to stream currents. Found among fallen leaves and debris gathered around stones in mid-stream, where the water is faster flowing.

Leonard Tan

Male, showing the characteristic yellow band on abdominal segment 7.

Erland Refling Nielsen

Female ovipositing in a roadside ditch at Fraser's Hill, Malaysia.

Leonard Tan

Lateral view of the male.

Leonard Tan

A female.

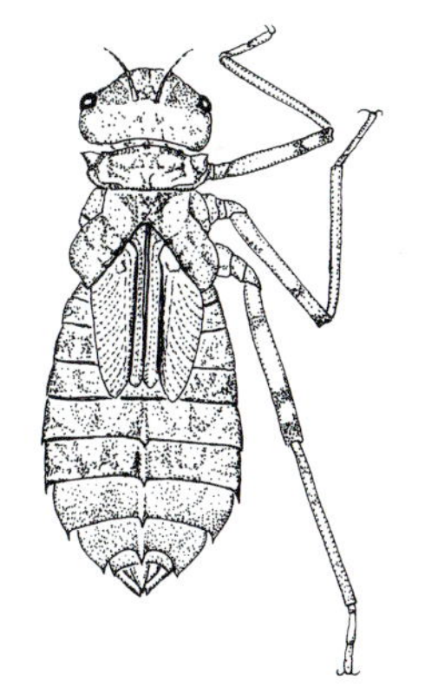

Drawing of the larva.

Synthemistidae (Tigertails)
This is a family of about 150 small to medium-sized dragonfly species of slender build, mostly from the southern hemisphere, Australasia being their centre of diversity. They are sometimes known as tigertails or southern emeralds, as some species are superficially similar to clubtails (Gomphidae) or emeralds (Corduliidae).

The family includes many dragonflies that were originally placed in the Corduliidae under various subfamilies, such as Gomphomacromiinae and Idionychinae. The family name comes from the genus *Synthemis*, which combines *syn-* (Greek for 'together with') and *-themis*, a common suffix for true dragonflies.

The status of the family remains uncertain and inconclusive, pending further molecular work to determine the relationships between its putative members. Until then, the family name is retained as it is the oldest available name in this complex of often fugitive and secretive dragonflies.

Tang Hung Bun

Male, showing the characteristic green eyes and extravagant anal appendages.

SHADOWDANCER *Idionyx yolanda* Selys, 1871

Size HWL: 27–29mm; TBL: 38–41mm

Description Small-medium, lightly built dragonfly with bright green eyes (brownish in younger individuals). Thorax dark metallic green with yellow stripe on sides. Abdomen thin and dark, expanding terminally, with very large appendages. Wings hyaline, with slight yellow tint at bases. Venation open, and anal loop present, though not sock shaped like those of emeralds and skimmers. Bases of male's hindwings slightly acute, but without sharp angle of cruisers. Sexes similar.

Habitat & Habits Found at small forest streams and shaded forest clearings. Rather fugitive species that has to be searched for in tangles of fallen branches or low vegetation near streams, where it perches in a hanging position. May be found 'dancing' on sunlit portions of forest trails near streams and swamps, but never too far from the shadows. Flight busy and bouncy, following small but erratic circuit that ranges from about knee height to a couple of metres above the ground. Sometimes a few individuals may fly a short distance from each other. Can be hard to follow as it flutters in and out of sunbeams, possibly chasing small dipterans, and easily vanishes from view when breaking off to settle in dense vegetation.

Presence in Singapore Recorded in forested habitats, including the Bukit Timah and Central Catchment Nature Reserves, adjacent nature parks and Bukit Brown. Described from a specimen collected in Singapore by Wallace.

Etymology Generic epithet combines the Greek terms for 'peculiar' or 'one's own' (*idio*) and 'claw' (*onyx*), referring to long teeth on tarsal claws. Yolanda (*Iolanthe* in Greek) is a female name that means 'violet'.

Distribution Sundaland.

National Conservation Status Least Concern; Widespread but Uncommon.

IUCN Red List Status Least Concern.

Larva Small (c. 12 mm) and overall very hairy, with divergent wing-buds and spidery legs. Can be quite numerous at breeding sites. Found among bottom detritus in streams.

Lena Chow

The Shadowdancer, the only member of its family to occur in Singapore.

Marcus Ng

Young male with brownish eyes.

Lee Yan Leong

Female with a yellow tint at the wing-bases.

Robin Ngiam

Dorsal view of the male's anal appendages.

CHECKLIST & NATIONAL STATUS OF SINGAPORE ODONATA

(Checklist compiled on 2/12/2021)

Abbreviations

CR Critically Endangered
EN Endangered
VU Vulnerable
NT Near Threatened
LC Least Concern
NE Nationally Extinct

Common Name	Species	Distribution and Rarity	Conservation Status
Suborder **Zygoptera**			
Family **Argiolestidae**			
Blue-spotted Flatwing	*Podolestes orientalis*	Restricted & Uncommon	VU
Family **Calopterygidae**			
White-faced Clearwing	*Echo modesta*	Nationally Extinct	NE
Green Metalwing	*Neurobasis chinensis*	Nationally Extinct	NE
Common Flashwing	*Vestalis amethystina*	Restricted but Common	VU
Charming Flashwing	*Vestalis amoena*	Restricted & Uncommon	EN
Plain Flashwing	*Vestalis gracilis*	Restricted but Common	CR
Family **Chlorocyphidae**			
Fiery Gem	*Libellago aurantiaca*	Restricted but Common	CR
Clearwing Gem	*Libellago hyalina*	Restricted but Common	CR
Golden Gem	*Libellago lineata*	Widespread & Common	LC
Orange-faced Gem	*Libellago stigmatizans*	Nationally Extinct	NE
Family **Coenagrionidae**			
Blue Slim	*Aciagrion hisopa*	Restricted & Very Rare	VU
Variable Wisp	*Agriocnemis femina*	Widespread & Common	LC
Marsh Wisp	*Agriocnemis minima*	Restricted & Very Rare	CR
Dwarf Wisp	*Agriocnemis nana*	Restricted & Very Rare	EN
Wandering Wisp	*Agriocnemis pygmaea*	Widespread but Rare	LC
Bebar Wisp	*Amphicnemis bebar*	Restricted & Very Rare	CR
Will-o-wisp	*Amphicnemis gracilis*	Restricted but Common	VU
Blue-nosed Sprite	*Archibasis melanocyana*	Restricted & Rare	EN
Rebecca's Sprite	*Archibasis rebeccae*	Restricted & Very Rare	CR
Violet Sprite	*Archibasis viola*	Widespread & Common	LC
Variable Sprite	*Argiocnemis rubescens rubeola*	Widespread but Uncommon	LC
Ornate Coraltail	*Ceriagrion cerinorubellum*	Widespread & Common	LC
Fiery Coraltail	*Ceriagrion chaoi*	Restricted & Rare	VU
Common Bluetail	*Ischnura senegalensis*	Widespread & Common	LC
Blue Midget	*Mortonagrion aborense*	Restricted & Rare	CR
Arthur's Midget	*Mortonagrion arthuri*	Restricted & Rare	VU
Hooked Midget	*Mortonagrion falcatum*	Restricted & Very Rare	CR
Dryad	*Pericnemis stictica*	Restricted & Rare	VU
Look-alike Sprite	*Pseudagrion australasiae*	Restricted & Uncommon	VU
Blue Sprite	*Pseudagrion microcephalum*	Widespread & Common	LC
Grey Sprite	*Pseudagrion pruinosum*	Restricted but Common	VU
Orange-faced Sprite	*Pseudagrion rubriceps*	Widespread but Rare	NT
Cryptic Shadesprite	*Teinobasis cryptica*	Restricted & Very Rare	CR
Red-tailed Sprite	*Teinobasis ruficollis*	Widespread but Rare	NT
Family **Devadattidae**			
Malayan Grisette	*Devadatta argyoides*	Restricted & Uncommon	EN
Family **Euphaeidae**			

Common Name	Species	Distribution and Rarity	Conservation Status
Black Velvetwing	*Dysphaea dimidiata*	Nationally Extinct	NE
Blue-sided Satinwing	*Euphaea impar*	Widespread & Common	LC
Family **Lestidae**			
Crenulated Spreadwing	*Lestes praemorsus*	Widespread but Uncommon	LC
Great Spreadwing	*Orolestes wallacei*	Nationally Extinct	NE
Slender Spreadwing	*Platylestes heterostylus*	Restricted & Very Rare	CR
Family **Platycnemididae**			
White-tailed Sylvan	*Coeliccia albicauda*	Restricted & Very Rare	CR
Twin-spotted Sylvan	*Coeliccia didyma*	Restricted & Very Rare	CR
Telephone Sylvan	*Coeliccia octogesima*	Restricted but Common	VU
Yellow Featherlegs	*Copera marginipes*	Widespread & Common	LC
Variable Featherlegs	*Copera vittata*	Restricted & Rare	VU
Shorttail	*Onychargia atrocyana*	Widespread but Uncommon	LC
Collared Threadtail	*Prodasineura collaris*	Restricted & Uncommon	VU
Orange-striped Threadtail	*Prodasineura humeralis*	Widespread & Common	LC
Interrupted Threadtail	*Prodasineura interrupta*	Restricted & Uncommon	CR
Crescent Threadtail	*Prodasineura notostigma*	Widespread & Common	LC
Family **Platystictidae**			
Singapore Shadowdamsel	*Drepanosticta quadrata*	Restricted but Common	VU
Suborder **Anisoptera**			
Family **Aeshnidae**			
Emperor	*Anax guttatus*	Widespread but Uncommon	LC
Arrow Emperor	*Anax panybeus*	Widespread but Rare	LC
Spoon-tailed Duskhawker	*Gynacantha basiguttata*	Restricted & Rare	VU
Small Duskhawker	*Gynacantha bayadera*	Widespread but Rare	LC
Spear-tailed Duskhawker	*Gynacantha dohrni*	Widespread but Uncommon	LC
Dingy Duskhawker	*Gynacantha subinterrupta*	Widespread but Uncommon	LC
Nighthawker	*Heliaeschna crassa*	Restricted & Very Rare	CR
Plain Nighthawker	*Heliaeschna simplicia*	Restricted & Very Rare	CR
Lesser Nighthawker	*Heliaeschna uninervulata*	Widespread but Rare	NT
Paddletail	*Oligoaeschna amata*	Restricted & Very Rare	CR
Leaftail	*Oligoaeschna foliacea*	Restricted & Very Rare	CR
Giant Hawker	*Tetracanthagyna plagiata*	Restricted & Uncommon	VU
Family **Corduliidae**			
Emerald	*Hemicordulia tenera*	Restricted & Rare	VU
Family **Gomphidae**			
Malayan Hooktail	*Acrogomphus malayanus*	Restricted & Rare	VU
Arthur's Clubtail	*Burmagomphus arthuri*	Restricted & Very Rare	CR
Splayed Clubtail	*Burmagomphus divaricatus*	Nationally Extinct	NE
Lesser Splayed Clubtail	*Burmagomphus plagiatus*	Nationally Extinct	NE
Malayan Grappletail	*Heliogomphus kelantanensis*	Restricted & Rare	CR
Common Flangetail	*Ictinogomphus decoratus*	Widespread & Common	LC
Ris' Clubtail	*Leptogomphus risi*	Restricted & Rare	VU
Forktail	*Macrogomphus quadratus*	Restricted & Uncommon	VU
Malayan Spineleg	*Merogomphus femoralis*	Restricted & Very Rare	CR
Tiny Sheartail	*Microgomphus chelifer*	Restricted & Rare	VU
Banded Hooktail	*Paragomphus capricornis*	Restricted & Rare	EN
Family **Libellulidae**			
Trumpet Tail	*Acisoma panorpoides*	Widespread & Common	LC
Blue Adjutant	*Aethriamanta aethra*	Widespread but Uncommon	LC
Scarlet Adjutant	*Aethriamanta brevipennis*	Widespread but Uncommon	LC
Pond Adjutant	*Aethriamanta gracilis*	Widespread & Common	LC
Grenadier	*Agrionoptera insignis*	Widespread and Common	LC
Handsome Grenadier	*Agrionoptera sexlineata*	Widespread but Uncommon	LC
Blue Dasher	*Brachydiplax chalybea*	Widespread & Common	LC
Black-tailed Dasher	*Brachydiplax farinosa*	Restricted & Very Rare	EN
Pixie	*Brachygonia oculata*	Restricted & Uncommon	EN

Common Name	Species	Distribution and Rarity	Conservation Status
Common Amberwing	*Brachythemis contaminata*	Widespread & Common	LC
Sultan	*Camacinia gigantea*	Widespread but Uncommon	LC
Green-eyed Percher	*Chalybiothemis fluviatilis*	Restricted & Uncommon	VU
Lined Forest Skimmer	*Cratilla lineata*	Widespread but Uncommon	LC
Dark-tipped Forest Skimmer	*Cratilla metallica*	Widespread & Common	LC
Common Scarlet	*Crocothemis servilia*	Widespread & Common	LC
Black-tipped Percher	*Diplacodes nebulosa*	Widespread but Uncommon	LC
Blue Percher	*Diplacodes trivialis*	Widespread & Common	LC
Water Monarch	*Hydrobasileus croceus*	Widespread & Common	LC
White-tipped Demon	*Indothemis carnatica*	Restricted & Very Rare	CR
Restless Demon	*Indothemis limbata*	Restricted & Uncommon	VU
Scarlet Grenadier	*Lathrecista asiatica*	Widespread & Common	LC
Bombardier	*Lyriothemis cleis*	Restricted & Rare	EN
Coastal Glider	*Macrodiplax cora*	Widespread & Common	LC
Scarlet Pygmy	*Nannophya pygmaea*	Widespread & Common	LC
Striped Grenadier	*Nesoxenia lineata*	Widespread but Uncommon	LC
Rare Parasol	*Neurothemis disparilis*	Nationally Extinct	NE
Common Parasol	*Neurothemis fluctuans*	Widespread & Common	LC
Riverhawk	*Onychothemis testacea*	Restricted & Very Rare	EN
Blue Sentinel	*Orchithemis pruinans*	Restricted & Rare	CR
Variable Sentinel	*Orchithemis pulcherrima*	Widespread & Common	LC
Spine-tufted Skimmer	*Orthetrum chrysis*	Widespread & Common	LC
Common Blue Skimmer	*Orthetrum glaucum*	Widespread & Common	LC
Slender Blue Skimmer	*Orthetrum luzonicum*	Widespread & Common	LC
Variegated Green Skimmer	*Orthetrum sabina*	Widespread & Common	LC
Scarlet Skimmer	*Orthetrum testaceum*	Widespread & Common	LC
Wandering Glider	*Pantala flavescens*	Widespread & Common	LC
Mangrove Marshal	*Pornothemis starrei*	Widespread but Uncommon	NT
Common Chaser	*Potamarcha congener*	Widespread & Common	LC
Banded Skimmer	*Pseudothemis jorina*	Widespread but Uncommon	LC
Mangrove Dwarf	*Raphismia bispina*	Widespread but Uncommon	NT
Common Redbolt	*Rhodothemis rufa*	Widespread & Common	LC
Small Bronze Flutterer	*Rhyothemis fulgens*	Nationally Extinct	NE
Bronze Flutterer	*Rhyothemis obsolescens*	Widespread but Uncommon	LC
Yellow-barred Flutterer	*Rhyothemis phyllis*	Widespread & Common	LC
Sapphire Flutterer	*Rhyothemis triangularis*	Widespread but Uncommon	LC
Potbellied Elf	*Risiophlebia dohrni*	Restricted & Rare	EN
Elf	*Tetrathemis hyalina*	Restricted & Very Rare	EN
White-barred Duskhawk	*Tholymis tillarga*	Widespread & Common	LC
Saddlebag Glider	*Tramea transmarina euryale*	Widespread & Common	LC
Crimson Dropwing	*Trithemis aurora*	Widespread & Common	LC
Indigo Dropwing	*Trithemis festiva*	Widespread & Common	LC
Dancing Dropwing	*Trithemis pallidinervis*	Widespread but Uncommon	LC
Treehugger	*Tyriobapta torrida*	Widespread & Common	LC
Rare Basker	*Urothemis abbotti*	Nationally Extinct	NE
Scarlet Basker	*Urothemis signata insignata*	Widespread & Common	LC
White Duskdarter	*Zyxomma obtusum*	Restricted & Uncommon	CR
Slender Duskdarter	*Zyxomma petiolatum*	Widespread & Common	LC
Family **Macromiidae**			
Pond Cruiser	*Epophthalmia vittigera*	Widespread & Common	LC
Stream Cruiser	*Macromia cincta*	Restricted & Rare	EN
Lesser Stream Cruiser	*Macromia cydippe*	Restricted & Rare	EN
Family **Synthemistidae**			
Shadowdancer	*Idionyx yolanda*	Widespread but Uncommon	LC

Selected References

Cai, Y., Ng, C. Y. & Ngiam, R. W. J. 2018. 'Diversity, distribution and habitat characteristics of dragonflies in Nee Soon freshwater swamp forest, Singapore'. *Gardens' Bulletin Singapore* 70 (Suppl. 1): 123–15370 (Suppl. 1): 123–153.

Cai, Y., Nga, Y. P. Q. & Ngiam, R. W. J. 2019. 'Diversity and Distribution of Dragonflies in Bukit Timah Nature Reserve, Singapore'. *Gardens' Bulletin Singapore* 71 (Suppl. 1): 293–316.

Cheong, L. F., Tang, H. B. & Ngiam, R. W. J. 2010. 'Ode to Odonata'. *NatureWatch* Jan–March 2010, 8–16.

Chow L., Gan, C. W. & Tsang, K. C. 2012. *A Field Guide to the Dragonflies of Singapore.* The Nature Society (Singapore), Singapore.

Corbet, P. S., 2004. *Dragonflies: Behaviour and Ecology of Odonata* (revised ed.). Harley Books, Colchester.

Davison, G. W. H., Ng, P. K. L. & Ho, H. C. (eds). 2008. *The Singapore Red Data Book: Threatened Plants and Animals of Singapore* (2nd ed.). The Nature Society (Singapore), Singapore.

Endersby, I. D. 2012. 'Etymology of the dragonflies (Insecta: Odonata) named by R. J. Tillyard, F.R.S.' *Proceedings of the Linnean Society of New South Wales* 134, 1–16.

Endersby, I. D. 2012. 'Watson and Theischinger: the etymology of the dragonfly (Insecta: Odonata) names which they published'. *Journal and Proceedings of the Royal Society of New South Wales*, vol. 145, nos 443 & 444, 34–53.

Fliedner, H. 2006. 'The scientific names of the Odonata in Burmeister's "Handbuch der Entomologie"'. *Virgo*, Die Verienszeitschrift der Entomologischer Verein Mecklenburg e.V., 9: 5–23.

Fliedner, H. 2020. 'The scientific names of Brauer's odonate taxa'. *IDF Report: Journal of the International Dragonfly Fund* 148: 1–55.

Fliedner, H. 2021. 'The scientific names of Ris' odonate taxa'. *IDF Report: Journal of the International Dragonfly Fund* 148: 1–146.

Fliedner, H. 2021. 'The scientific names of Krüger's odonate taxa with annotations about his contribution to neuropterological taxonomy'. *IDF Report: Journal of the International Dragonfly Fund* 167, 1–62.

Karjalainen, S. & Hämäläinen, M. 2013. *Demoiselle Damselflies: Winged Jewels of Silvery Streams.* Caloptera Publishing, Helsinki.

Laidlaw, F. F. 1902. 'On a collection of dragonflies made by Members of the Skeat Expedition in the Malay Peninsula in 1899–1900'. *Proceedings of the General Meetings for Scientific Business of the Zoological Society of London*, 1 (January–April): 63–92.

Laidlaw, F. F. 1931. 'A list of the dragonflies (Odonata) of the Malay Peninsula with descriptions of new species'. *Journal of the Federated Malay States Museum* 16: 175–233.

Lieftinck, M. A. 1954. 'Handlist of Malaysian Odonata: a catalogue of the dragonflies of the Malay Peninsula, Sumatra, Java and Borneo, including the adjacent small islands'. *Treubia* 22 (Suppl.).

Murphy, D. H. 1997. 'Odonata biodiversity in the nature reserves of Singapore'. *Gardens' Bulletin, Singapore* 49: 333–352.

Ng, M. 2020. 'Finding dragonflies in Singapore's nature parks'. *Agrion* 24(2), May 2020: 66–68.

Ngiam, R. W. J. 2011. *Dragonflies of Our Parks and Gardens.* National Parks Board, Singapore.

Ngiam R. W. J. & Davison G. W. H. 2011. 'A checklist of dragonflies in Singapore parks (Odonata: Anisoptera, Zygoptera)'. *Nature in Singapore* 4: 349–353.

Ngiam, R. W. J. & Cheong, L. F. 2016. 'The dragonflies of Singapore: an updated checklist and revision of the national conservation statuses'. *Nature in Singapore* 9: 149–163.

Norma-Rashid, Y., Cheong, L. F., Lua, H. K. & Murphy, D. H. 2008. *The Dragonflies (Odonata) of Singapore: Current Status Records and Collections of the Raffles Museum of Biodiversity Research.* Raffles Museum of Biodiversity Research, Singapore.

Orr, A. G. 2001. 'An annotated checklist of the Odonata of Brunei with ecological notes and descriptions of hitherto unknown males and larvae'. *International Journal of Odonatology* 4 (2): 167–220.

Orr, A. G. 2003. *A Guide to the Dragonflies of Borneo: Their Identification and Biology.* Natural History Publications (Borneo), Kota Kinabalu.

Orr, A. G. 2005. *Dragonflies of Peninsular Malaysia and Singapore.* Natural History Publications (Borneo), Kota Kinabalu.

Orr, A. G. & Hämäläinen, M. 2007. *The Metalwing Demoiselles of the Eastern Tropics: Their Identification and Biology.* Natural History Publications (Borneo), Kota Kinabalu.

Paulson, D. 2019. *Dragonflies & Damselflies: A Natural History.* Ivy Press, Brighton.

Silsby, J. 2001. *Dragonflies of the World.* CSIRO Publishing, Collingwood.

Soh, M., Ng, M. & Ngiam, R. W. J. 2019. 'New Singapore record of a dragonfly, *Indothemis carnatica*, with an updated Singapore Odonata checklist'. *Singapore Biodiversity Records* 2019: 10–17.

Tang, H. B., Wang, L. K. & Hämäläinen, M. 2010. *A Photographic Guide to the Dragonflies of Singapore.* Raffles Museum of Biodiversity Research, Singapore.

Yokoi, N. 1995. 'A record of the Odonata from Mandai, Singapore'. *Malangpo* (Newsletter of the Thai National Office of the International Odonatological Society), 13: 100.

Dragonfly Websites

http://addo.adu.org.za/index.php 'ADDO' (African Dragonflies and Damselflies Online). Much useful information on odonate anatomy as well as genera found in Asia as well as Africa.

http://dragonflyfund.org/en Website of the International Dragonfly Fund, containing links to their journals *IDF-Report* and *Faunistic Studies in South-East Asian and Pacific Island Odonata*.

https://worlddragonfly.org Website of the Worldwide Dragonfly Association, including its highly informative bulletin *Agrion*.

https://www.odonatacentral.org/app/#/wol 'World Odonata List' coordinated by Dennis Paulson et al.

https://singaporeodonata.wordpress.com 'Dragonflies & Damselflies of Singapore' by Anthony Quek.

https://sgodonata.wordpress.com 'Picture of Singapore Odonata' by Leonard Tan.

www.facebook.com/groups/295432323905656 'Dragonflies of Singapore' Facebook group by Lena Chow and Ng Soon Chye.

http://odonata-malaysia.blogspot.com 'Odonata of Peninsular Malaysia' by C.Y. Choong.

http://thaiodonata.blogspot.com 'Dragonflies & Damselflies of Thailand' by Dennis Farrell.

http://odonatavietnam.blogspot.com 'Dragonflies & Damselflies of Vietnam' by Tom Kompier.

https://atratothemis.com 'Dragonflies of Hong Kong, Hainan and Sarawak' by Graham Reels.

www.biosch.hku.hk/ecology/staffhp/dd/macroinvertebrates/Odonata/odonata.html Freshwater Macroinvertebrates in Hong Kong: odonate larvae.

https://lkcnhm.nus.edu.sg/publications/nature-in-singapore/volumes The *Nature of Singapore* and *Singapore Biodiversity Records* online journals, containing various papers on the habitats and ecology, as well as sightings records, of local dragonflies.

Acknowledgements

Tang Hung Bun for his encouragement, enthusiasm and generous sharing of photographs. Lee Kong Chian Natural History Museum for access to specimens, QR code links and permission to reproduce and adapt the glossary from Tang et al. (2010). Special thanks to Rory Dow and Albert Orr for taxonomic advice, knowledge sharing and illustrations. Geoffrey Davison connected the authors to the publisher.

All the photographers who generously shared their images regardless of whether they were eventually selected: Antonio Giudici, Anthony Quek, Bennett Tan, Benoît Guillon, Bill Ho (Hong Kong AFCD), Bo Nielsen, Catalina Tong, Cheong Loong Fah, Chien Lee, Choong Chee Yen, Chris Ang, Damian Pinguey, Dennis Farrell, Erland Refling Nielsen, Eugene Tay, Fiora Li, Graham Reels, Guek Hock Ping (Kurt), Gunther Theischinger, Henrietta Woo, James Holden, Justin Tan, Keith Wilson, Kenneth Fong, Laurence Leong, Lena Chow, Leonard Tan, Leslie Day, Mandy Lee, Matti Hämäläinen, Marcel Finlay, Marcel Silvius, Martin Kennewell, Max Khoo, Meena Vathyam, Michael Soh, Oleg Kosterin, Phil Benstead, QM Zhou, Reagan Villaneuva, Rory Dow, Spencer Yau, Tan Hui Zhen, Tom Kompier, Veronica Foo Tse Fen, Wu Hongdao (via Zhang Haomiao), Yeo Suay Hwee, Yue Teng Lee.

Lastly to the odonates, whose very existence is a source of delight!

SPECIES INDEX